Spring Boot 区块链
应用开发入门 微课视频版

吴 胜 ◎ 编著

清华大学出版社

北京

内 容 简 介

本书先介绍区块链基础知识和 Spring Boot 开发基础知识，如何用 Spring Boot 实现区块链应用的基础开发和 P2P 的实现，消息机制的实现，数据库开发，共识算法的实现，工作流、多语言和智能合约的应用等内容；然后介绍如何应用几款常用的用 Java 开发的区块链库、API 和框架；最后，结合三个简单案例演示 Spring Boot 区块链应用开发的过程。

本书结合示例，由浅入深地介绍各个知识点，并按照区块链应用开发技术所处层次由低到高、Spring Boot 开发由易到难和开发步骤先后次序来组织章节内容，还配备了示例的操作视频，可以帮助读者更好地掌握如何用 Spring Boot 进行区块链应用开发。

本书适合 Spring Boot 开发和区块链应用开发的初学者(特别是在校学生)、Web 应用开发者、企业级应用开发者等读者，还可以作为高等院校相关专业的教材、自学的入门读物、学习过程的参考书。

本书封面贴有清华大学出版社防伪标签，无标签者不得销售。
版权所有，侵权必究。举报：010-62782989，beiqinquan@tup.tsinghua.edu.cn。

图书在版编目(CIP)数据

Spring Boot 区块链应用开发入门：微课视频版/吴胜编著.—北京：清华大学出版社，2020.5(2024.8重印)
(清华科技大讲堂)
ISBN 978-7-302-55257-4

Ⅰ.①S… Ⅱ.①吴… Ⅲ.①JAVA 语言－程序设计 Ⅳ.①TP312.8

中国版本图书馆 CIP 数据核字(2020)第 051944 号

责任编辑：陈景辉　张爱华
封面设计：刘　键
责任校对：时翠兰
责任印制：刘　菲

出版发行：清华大学出版社
　　　　网　　址：https://www.tup.com.cn，https://www.wqxuetang.com
　　　　地　　址：北京清华大学学研大厦 A 座　　　邮　编：100084
　　　　社 总 机：010-83470000　　　　　　　　　　邮　购：010-62786544
　　　　投稿与读者服务：010-62776969，c-service@tup.tsinghua.edu.cn
　　　　质量反馈：010-62772015，zhiliang@tup.tsinghua.edu.cn
　　　　课件下载：https://www.tup.com.cn，010-83470236
印 装 者：三河市君旺印务有限公司
经　　销：全国新华书店
开　　本：185mm×260mm　　印　张：22　　　　　字　数：536 千字
版　　次：2020 年 7 月第 1 版　　　　　　　　　　印　次：2024 年 8 月第 3 次印刷
印　　数：2501～2600
定　　价：69.90 元

产品编号：084753-01

前 言

区块链技术是以比特币为代表的数字加密货币体系的核心支撑技术,具有高度透明、去中心化、去信任、集体维护、匿名等特性,能够通过运用数据加密、时间戳、分布式共识和经济激励等手段,在节点无须互相信任的分布式系统中实现基于去中心化信用的点对点交易、协调与协作,从而为解决中心化机构普遍存在的高成本、低效率和数据存储不安全等问题提供了解决方案。区块链技术作为下一代互联网的核心技术之一,其应用实现技术与分布式系统(微服务架构)、NoSQL 数据库、消息机制、工作流、多语言开发等热门技术密切相关,同时与大数据、人工智能等技术融合,可向众多软件系统提供基础性支持,具有广阔的应用前景。区块链技术的应用将会给金融等众多领域带来翻天覆地式的变化,还可以被广泛应用于供应链、政务、慈善等众多领域。

基于 Spring Boot 具备简单、易用、易学、易上手等特点,本书以 Spring Boot 进行区块链应用开发。

考虑到初学者(特别是在校学生)对分布式系统、Spring Boot 和区块链应用的开发经验较少,为了帮助读者更好地安排学习时间和教师更好地安排教学进度,给出如下表所述的各章的建议学时。

章 内 容	建议理论学时	建议实践学时
第 1 章　区块链基础	2	0.5
第 2 章　Spring Boot 开发基础	2	1
第 3 章　区块链应用的基础开发	3.5	2.5
第 4 章　区块链应用的 P2P 实现	3.5	3
第 5 章　区块链应用中消息机制的实现	2.5	2.5
第 6 章　区块链应用中数据库开发	4	3
第 7 章　区块链应用中共识算法的实现	3	2
第 8 章　工作流、多语言和智能合约的应用	2.5	2.5
第 9 章　区块链库、API 和框架的应用	3.5	3.5
第 10 章　基于区块链的简易系统开发	3.5	3.5
合计学时	30	24

在开设 Spring Boot 区块链应用开发相关课程时,教师可以根据实际情况进行调整。读者也可以有选择地阅读章节内容、安排学时。

本书特色

(1) 以 Spring Boot 为开发工具,为 Java 开发人员快速了解区块链开发基础知识提供了便利。

(2) 实战案例丰富,包括 37 个知识点案例和 3 个完整项目案例。

(3) 内容由浅入深、循序渐进,代码详尽。

配套资源

为便于教与学,本书配有全书源代码、教学课件、教学大纲、教案、教学进度表、习题参考答案及 222 分钟微课视频。

(1) 获取教学视频方式:读者可以先扫描本书封底的文泉云盘防盗码,再扫描书中相应的视频二维码,观看教学视频。

(2) 获取源代码、习题参考答案:先扫描本书封底的文泉云盘防盗码,再扫描下方二维码,即可获取。

源代码

关于如何直接使用源代码的说明

习题参考答案

(3) 其他配套资源可以扫描本书封底的课件二维码下载。

读者对象

本书主要面向广大从事区块链开发、Java 开发的专业人员,从事高等教育的教师,高等院校的在读学生及相关领域的广大科研人员。

本书的编写参考了诸多相关资料,在此表示衷心的感谢。限于个人水平和时间仓促,书中难免存在疏漏之处,欢迎读者批评指正。

编者

2020 年 7 月

目 录

第 1 章 区块链基础 ... 1

1.1 区块链简介 ... 1
- 1.1.1 区块链的含义 ... 1
- 1.1.2 区块链的特点 ... 2
- 1.1.3 区块链的分类 ... 2

1.2 区块链的发展 ... 3
- 1.2.1 区块链的发展简史 ... 3
- 1.2.2 区块链的发展阶段 ... 3

1.3 区块链的应用领域 ... 4
- 1.3.1 信息数据领域 ... 4
- 1.3.2 金融与数字资产领域 ... 4
- 1.3.3 供应链领域 ... 5
- 1.3.4 政务领域 ... 5

1.4 区块链应用开发的相关技术 ... 5
- 1.4.1 技术模型 ... 5
- 1.4.2 支持环境 ... 5
- 1.4.3 信任机制 ... 6
- 1.4.4 智能合约 ... 6
- 1.4.5 应用系统 ... 6
- 1.4.6 区块链技术的多样性 ... 7

1.5 比特币钱包的安装与使用 ... 7
- 1.5.1 Electrum 钱包的功能 ... 7
- 1.5.2 网页版 Coin.Space 钱包的功能 ... 9

习题 1 ... 10

第 2 章 Spring Boot 开发基础 ... 11

2.1 Spring Boot 基础 ... 11
- 2.1.1 Spring Boot 简介 ... 11
- 2.1.2 Spring Boot 的主要特点 ... 11

2.2 配置 Spring Boot 开发环境 ... 12

2.2.1　安装和配置 JDK ……………………………………………………… 12
　　　2.2.2　安装 IDEA …………………………………………………………… 12
　2.3　创建项目与项目基本情况 …………………………………………………………… 13
　　　2.3.1　利用 IDEA 创建项目 ………………………………………………… 13
　　　2.3.2　项目的基本构成 ……………………………………………………… 16
　2.4　Spring Boot 开发起步 ……………………………………………………………… 17
　　　2.4.1　软件生命周期 ………………………………………………………… 17
　　　2.4.2　Spring Boot 开发步骤 ………………………………………………… 17
　2.5　helloworld 项目的实现 ……………………………………………………………… 18
　　　2.5.1　添加依赖 ……………………………………………………………… 18
　　　2.5.2　创建类 Block ………………………………………………………… 18
　　　2.5.3　创建类 SecurityUtils ………………………………………………… 19
　　　2.5.4　创建类 HelloController ……………………………………………… 20
　　　2.5.5　运行程序 ……………………………………………………………… 21
习题 2 …………………………………………………………………………………………… 22

第 3 章　区块链应用的基础开发 …………………………………………………………… 23

　3.1　模拟两个钱包之间的交互 …………………………………………………………… 23
　　　3.1.1　创建项目并添加依赖 ………………………………………………… 23
　　　3.1.2　创建接口 MainService ……………………………………………… 23
　　　3.1.3　创建类 MainServiceImpl …………………………………………… 24
　　　3.1.4　创建类 MainController ……………………………………………… 25
　　　3.1.5　修改配置文件 application.properties ……………………………… 26
　　　3.1.6　运行程序 ……………………………………………………………… 26
　3.2　模拟挖矿 ……………………………………………………………………………… 28
　　　3.2.1　添加依赖 ……………………………………………………………… 28
　　　3.2.2　创建类 Block ………………………………………………………… 28
　　　3.2.3　创建类 MineService ………………………………………………… 29
　　　3.2.4　创建类 MineController ……………………………………………… 31
　　　3.2.5　运行程序 ……………………………………………………………… 31
　3.3　模拟创建钱包和利用钱包进行交易 ………………………………………………… 32
　　　3.3.1　添加依赖 ……………………………………………………………… 32
　　　3.3.2　创建类 StringUtil …………………………………………………… 32
　　　3.3.3　创建类 ContractService ……………………………………………… 34
　　　3.3.4　创建类 ErcController ………………………………………………… 35
　　　3.3.5　运行程序 ……………………………………………………………… 36
　3.4　Merkle 树的实现 …………………………………………………………………… 37
　　　3.4.1　创建类 Stakeholder ………………………………………………… 37
　　　3.4.2　创建类 ProofEntry …………………………………………………… 38

  3.4.3 创建类 Node ····· 38
  3.4.4 创建类 FtsResult ····· 39
  3.4.5 创建类 FtsService ····· 40
  3.4.6 创建类 FtsController ····· 42
  3.4.7 运行程序 ····· 43
 3.5 Gossip 协议的模拟实现 ····· 44
  3.5.1 创建项目并添加依赖 ····· 44
  3.5.2 创建类 Node ····· 44
  3.5.3 创建类 StaticSeedFinder ····· 45
  3.5.4 创建类 SeedFinderChain ····· 45
  3.5.5 创建类 GossipConnector ····· 46
  3.5.6 创建类 GossipController ····· 47
  3.5.7 运行程序 ····· 48
 3.6 模拟 IOTA 的应用 ····· 48
  3.6.1 创建项目并添加依赖 ····· 48
  3.6.2 创建类 IotaController ····· 49
  3.6.3 运行程序 ····· 49
 3.7 用线程模拟区块链的示例 ····· 50
  3.7.1 创建项目并添加依赖 ····· 50
  3.7.2 创建类 Block ····· 50
  3.7.3 创建类 StringUtil ····· 51
  3.7.4 创建类 MineThread ····· 52
  3.7.5 创建类 OutWaitThread ····· 52
  3.7.6 创建类 MineController ····· 53
  3.7.7 创建类 BlockController ····· 54
  3.7.8 创建类 BlockchainController ····· 55
  3.7.9 运行程序 ····· 56
 习题 3 ····· 58

第 4 章 区块链应用的 P2P 实现

 4.1 基于 Java-WebSocket 实现 P2P 网络 ····· 59
  4.1.1 创建项目并添加依赖 ····· 59
  4.1.2 创建类 P2PUtil ····· 60
  4.1.3 创建类 P2Pserver ····· 60
  4.1.4 创建类 P2Pclient ····· 61
  4.1.5 运行程序 ····· 62
 4.2 基于 WebSocket 实现 P2P 页面互连 ····· 63
  4.2.1 添加依赖 ····· 63
  4.2.2 创建类 WalletSendMessage ····· 63

4.2.3　创建类 WebSocketConfig ……………………………………………… 64
　　　4.2.4　创建类 WalletController ……………………………………………… 64
　　　4.2.5　创建文件 index.html ………………………………………………… 65
　　　4.2.6　运行程序 ……………………………………………………………… 67
　4.3　基于 t-io 实现 P2P 网络 …………………………………………………………… 68
　　　4.3.1　添加依赖 ……………………………………………………………… 69
　　　4.3.2　创建类 ServerPacket ………………………………………………… 69
　　　4.3.3　创建类 MServerAioHandler …………………………………………… 69
　　　4.3.4　创建类 MServerAioListener …………………………………………… 71
　　　4.3.5　创建类 MClientAioHandler …………………………………………… 72
　　　4.3.6　创建类 MClientAioListener …………………………………………… 73
　　　4.3.7　创建类 TIOServer ……………………………………………………… 74
　　　4.3.8　创建类 TIOClient ……………………………………………………… 75
　　　4.3.9　创建类 TIOController ………………………………………………… 76
　　　4.3.10　运行程序 ……………………………………………………………… 76
　4.4　基于 ZooKeeper 实现 P2P 服务 …………………………………………………… 78
　　　4.4.1　服务提供者模块 provider 的实现 ……………………………………… 78
　　　4.4.2　消费者模块 consumer 的实现 ………………………………………… 81
　　　4.4.3　运行程序 ……………………………………………………………… 82
　4.5　基于 Web Service 和 CXF 实现 P2P 服务 ………………………………………… 83
　　　4.5.1　服务器端模块 serverofws 的实现 ……………………………………… 83
　　　4.5.2　客户端模块 clientofws 的实现 ………………………………………… 85
　　　4.5.3　运行程序 ……………………………………………………………… 86
　4.6　同一服务器向多个页面发送区块链信息的示例 …………………………………… 87
　　　4.6.1　创建项目并添加依赖 …………………………………………………… 87
　　　4.6.2　创建类 Block …………………………………………………………… 87
　　　4.6.3　创建类 SocketIOConfig ………………………………………………… 88
　　　4.6.4　创建类 Service ………………………………………………………… 89
　　　4.6.5　创建类 ServerRunner ………………………………………………… 90
　　　4.6.6　创建类 MsgEventHandler ……………………………………………… 91
　　　4.6.7　创建类 ClientController ………………………………………………… 92
　　　4.6.8　创建类 StringUtil ……………………………………………………… 93
　　　4.6.9　创建类 LoggerUtil ……………………………………………………… 94
　　　4.6.10　创建文件 index.html、login.html 和 welcome.html …………………… 94
　　　4.6.11　修改配置文件 application.properties ………………………………… 96
　　　4.6.12　运行程序 ……………………………………………………………… 96
习题 4 ……………………………………………………………………………………… 98

第 5 章 区块链应用中消息机制的实现 ………………………………………………… 99

5.1 ActiveMQ 的应用 ………………………………………………… 99
5.1.1 创建项目并添加依赖 ………………………………………………… 99
5.1.2 创建类 Sender ………………………………………………… 100
5.1.3 创建类 Receiver ………………………………………………… 101
5.1.4 创建类 RsUtil ………………………………………………… 102
5.1.5 创建类 ActiveMQController ………………………………………………… 102
5.1.6 运行程序 ………………………………………………… 103

5.2 RabbitMQ 的应用 ………………………………………………… 104
5.2.1 创建项目并添加依赖 ………………………………………………… 104
5.2.2 创建类 Runner ………………………………………………… 104
5.2.3 创建类 Receiver ………………………………………………… 105
5.2.4 修改入口类 ………………………………………………… 106
5.2.5 运行程序 ………………………………………………… 107

5.3 Spring Cloud Stream 和 RabbitMQ 的应用 ………………………………………………… 108
5.3.1 创建项目并添加依赖 ………………………………………………… 108
5.3.2 创建接口 Sink ………………………………………………… 110
5.3.3 创建类 SinkReceiver ………………………………………………… 110
5.3.4 创建配置文件 application.yml ………………………………………………… 111
5.3.5 运行程序 ………………………………………………… 111

5.4 基于 ActiveMQ 传递区块链消息的示例 ………………………………………………… 112
5.4.1 创建项目并添加依赖 ………………………………………………… 112
5.4.2 创建类 Block ………………………………………………… 112
5.4.3 创建类 MQSendService ………………………………………………… 113
5.4.4 创建类 MQReceiveService ………………………………………………… 114
5.4.5 创建类 StringUtil ………………………………………………… 114
5.4.6 创建类 SendInfoController ………………………………………………… 115
5.4.7 修改配置文件 application.properties ………………………………………………… 116
5.4.8 运行程序 ………………………………………………… 116

习题 5 ………………………………………………… 117

第 6 章 区块链应用中数据库开发 ………………………………………………… 118

6.1 MySQL 的应用 ………………………………………………… 118
6.1.1 创建项目并添加依赖 ………………………………………………… 118
6.1.2 创建类 Pair ………………………………………………… 119
6.1.3 创建接口 PairService ………………………………………………… 120
6.1.4 创建类 PairServiceImpl ………………………………………………… 120
6.1.5 创建接口 PairRepository ………………………………………………… 122

6.1.6　创建类 PairController ……………………………………………… 122
　　6.1.7　修改配置文件 application.properties ……………………………… 123
　　6.1.8　运行程序 …………………………………………………………… 123
6.2　CouchDB 的应用 ………………………………………………………………… 124
　　6.2.1　创建项目并添加依赖 ………………………………………………… 124
　　6.2.2　创建类 CouchDBConfiguration ……………………………………… 125
　　6.2.3　创建类 Note ………………………………………………………… 126
　　6.2.4　创建类 NotFoundException ………………………………………… 127
　　6.2.5　创建类 NotePersistenceHandler ……………………………………… 127
　　6.2.6　创建类 NoteDTO ……………………………………………………… 128
　　6.2.7　创建类 NoteService …………………………………………………… 128
　　6.2.8　创建类 NotesController ……………………………………………… 129
　　6.2.9　修改配置文件 application.properties ……………………………… 130
　　6.2.10　修改入口类 ………………………………………………………… 130
　　6.2.11　运行程序 …………………………………………………………… 131
6.3　MongoDB 的应用 ………………………………………………………………… 132
　　6.3.1　创建项目并添加依赖 ………………………………………………… 132
　　6.3.2　创建类 Block ………………………………………………………… 133
　　6.3.3　创建接口 BlockRepository …………………………………………… 133
　　6.3.4　创建类 SHA256 ……………………………………………………… 134
　　6.3.5　创建类 PrimaryController …………………………………………… 134
　　6.3.6　修改配置文件 application.properties ……………………………… 135
　　6.3.7　运行程序 …………………………………………………………… 135
6.4　用以太坊区块链进行数据审核的示例 …………………………………………… 136
　　6.4.1　创建项目并添加依赖 ………………………………………………… 136
　　6.4.2　创建类 User …………………………………………………………… 136
　　6.4.3　创建类 Property ……………………………………………………… 138
　　6.4.4　创建类 Feedback ……………………………………………………… 139
　　6.4.5　创建类 ContractDetails ……………………………………………… 140
　　6.4.6　创建接口 Repository ………………………………………………… 141
　　6.4.7　创建接口 UserRepository …………………………………………… 141
　　6.4.8　创建接口 PropertyRepository ………………………………………… 142
　　6.4.9　创建接口 FeedbackRepository ……………………………………… 142
　　6.4.10　创建接口 EthereumContractRepository …………………………… 142
　　6.4.11　创建类 AbstractService ……………………………………………… 143
　　6.4.12　创建类 UserService ………………………………………………… 143
　　6.4.13　创建类 PropertyService ……………………………………………… 144
　　6.4.14　创建类 FeedbackService …………………………………………… 145
　　6.4.15　创建类 EthereumService …………………………………………… 146

- 6.4.16 创建类 ControllerLoggingAspect ·········· 148
- 6.4.17 创建类 Sha256Hex ·········· 149
- 6.4.18 创建类 HashArray ·········· 149
- 6.4.19 创建类 ApplicationConfig ·········· 151
- 6.4.20 创建类 UserController ·········· 152
- 6.4.21 创建类 PropertyController ·········· 154
- 6.4.22 创建类 FeedbackController ·········· 155
- 6.4.23 创建类 EthereumController ·········· 156
- 6.4.24 修改配置文件 application.properties ·········· 157
- 6.4.25 运行程序 ·········· 158

习题 6 ·········· 159

第 7 章 区块链应用中共识算法的实现 ·········· 160

7.1 PoW 算法的实现 ·········· 160
- 7.1.1 创建项目并添加依赖 ·········· 160
- 7.1.2 创建类 BeanInjector ·········· 161
- 7.1.3 创建类 BlockChain ·········· 162
- 7.1.4 创建类 RegisterRequest ·········· 165
- 7.1.5 创建类 Transaction ·········· 166
- 7.1.6 创建类 FastJsonUtil ·········· 166
- 7.1.7 创建类 SHAUtils ·········· 167
- 7.1.8 创建类 SwaggerConfig ·········· 168
- 7.1.9 创建类 BlockChainController ·········· 169
- 7.1.10 修改配置文件 application.properties ·········· 171
- 7.1.11 创建文件 index.html ·········· 172
- 7.1.12 运行程序 ·········· 172

7.2 PBFT 算法的实现 ·········· 174
- 7.2.1 创建项目并添加依赖 ·········· 174
- 7.2.2 创建类 PbftMsg ·········· 174
- 7.2.3 创建类 Pbft ·········· 175
- 7.2.4 创建类 PoAUtil ·········· 183
- 7.2.5 创建类 TimerManager ·········· 183
- 7.2.6 创建类 PbftController ·········· 184
- 7.2.7 运行程序 ·········· 185

7.3 Raft 算法的实现 ·········· 188
- 7.3.1 创建项目并添加依赖 ·········· 189
- 7.3.2 创建类 Follower ·········· 189
- 7.3.3 创建类 Candidate ·········· 190
- 7.3.4 创建类 Leader ·········· 193

7.3.5 创建类 ClusterMsg ……………………………………………………… 194
7.3.6 创建类 MsgUtil …………………………………………………………… 195
7.3.7 创建类 RaftController …………………………………………………… 196
7.3.8 修改配置文件 application.properties …………………………………… 196
7.3.9 运行程序 ………………………………………………………………… 197
7.4 基于 PoW 的区块链应用示例 ………………………………………………… 197
7.4.1 创建项目并添加依赖 …………………………………………………… 197
7.4.2 创建类 BaseEntity ……………………………………………………… 198
7.4.3 创建类 Block …………………………………………………………… 199
7.4.4 创建类 Blockchain ……………………………………………………… 199
7.4.5 创建类 Member ………………………………………………………… 200
7.4.6 创建类 MemberGroup …………………………………………………… 201
7.4.7 创建接口 MemberRepository …………………………………………… 201
7.4.8 创建接口 MemberGroupRepository ……………………………………… 201
7.4.9 创建接口 BlockchainRepository ………………………………………… 202
7.4.10 创建类 MemberService ………………………………………………… 202
7.4.11 创建类 MemberGroupService …………………………………………… 202
7.4.12 创建类 BlockchainService ……………………………………………… 203
7.4.13 创建类 ByteUtils ………………………………………………………… 204
7.4.14 创建类 ProofOfWork …………………………………………………… 204
7.4.15 创建类 PowResult ……………………………………………………… 206
7.4.16 创建类 MemberandGroupController …………………………………… 206
7.4.17 创建配置文件 application.yml ………………………………………… 207
7.4.18 运行程序 ………………………………………………………………… 207
习题 7 ……………………………………………………………………………………… 208

第 8 章 工作流、多语言和智能合约的应用 ……………………………………………… 209

8.1 Activiti 的应用 ……………………………………………………………………… 209
8.1.1 创建项目并添加依赖 …………………………………………………… 209
8.1.2 创建类 Applicant ………………………………………………………… 210
8.1.3 创建接口 ApplicantRepository ………………………………………… 211
8.1.4 创建类 ResumeService ………………………………………………… 211
8.1.5 创建类 HireProcessRestController ……………………………………… 211
8.1.6 修改配置文件 application.properties …………………………………… 212
8.1.7 修改入口类 ……………………………………………………………… 212
8.1.8 修改测试类 ……………………………………………………………… 213
8.1.9 运行程序 ………………………………………………………………… 215
8.2 Spring Cloud Sidecar 的多语言应用 ………………………………………………… 216
8.2.1 创建项目并添加依赖 …………………………………………………… 216

        8.2.2　修改配置文件 application.properties 218
        8.2.3　修改入口类 218
        8.2.4　创建文件 node-service.js 219
        8.2.5　Spring Cloud Eureka 注册中心的实现 219
        8.2.6　运行程序 222
    8.3　智能合约的模拟实现 223
        8.3.1　创建项目并添加依赖 223
        8.3.2　创建类 AccountNew 223
        8.3.3　创建接口 AccountRepository 224
        8.3.4　创建类 SmartContractStub 224
        8.3.5　创建接口 ISmartContract 225
        8.3.6　创建类 SCController 225
        8.3.7　修改配置文件 application.properties 228
        8.3.8　运行程序 228
    8.4　基于 Activiti 的区块链应用示例 229
        8.4.1　创建项目并添加依赖 229
        8.4.2　创建类 Person 230
        8.4.3　创建类 Comp 231
        8.4.4　创建类 Block 231
        8.4.5　创建类 Blockchain 232
        8.4.6　创建类 TaskRepresentation 232
        8.4.7　创建接口 PersonRepository 232
        8.4.8　创建接口 CompRepository 233
        8.4.9　创建接口 BlockchainRepository 233
        8.4.10　创建类 ActiveService 233
        8.4.11　创建类 JoinService 234
        8.4.12　创建类 MyRestController 236
        8.4.13　创建文件 join.bpmn20.xml 237
        8.4.14　修改配置文件 application.properties 239
        8.4.15　修改入口类 239
        8.4.16　运行程序 241
    习题 8 242

第 9 章　区块链库、API 和框架的应用 243

    9.1　bitcoinj 的应用 243
        9.1.1　bitcoinj 简介 243
        9.1.2　创建项目并添加依赖 244
        9.1.3　创建类 BitcoinJController 244
        9.1.4　运行程序 245

9.2 fabric-sdk-java 的应用 ··············· 245
　9.2.1 fabric-sdk-java 简介 ··············· 245
　9.2.2 添加依赖 ··············· 246
　9.2.3 创建类 HyperledgerController ··············· 246
　9.2.4 创建配置文件 config.properties ··············· 247
　9.2.5 运行程序 ··············· 247
9.3 eth-contract-api 的应用 ··············· 248
　9.3.1 添加依赖 ··············· 248
　9.3.2 创建类 EthcontractapiController ··············· 248
　9.3.3 运行程序 ··············· 249
9.4 exonum-java-binding 的应用 ··············· 250
　9.4.1 添加依赖 ··············· 250
　9.4.2 创建类 ExonumController ··············· 250
　9.4.3 运行程序 ··············· 251
9.5 web3j 的应用 ··············· 251
　9.5.1 web3j 简介 ··············· 251
　9.5.2 添加依赖 ··············· 252
　9.5.3 创建类 Web3jController ··············· 252
　9.5.4 运行程序 ··············· 253
9.6 WavesJ 的应用 ··············· 253
　9.6.1 添加依赖 ··············· 253
　9.6.2 创建类 WavesJController ··············· 254
　9.6.3 运行程序 ··············· 254
9.7 基于 web3j 钱包业务功能的示例 ··············· 255
　9.7.1 创建项目并添加依赖 ··············· 255
　9.7.2 创建类 BlockchainTransaction ··············· 255
　9.7.3 创建接口 BTxRepository ··············· 256
　9.7.4 创建类 BlockchainService ··············· 256
　9.7.5 创建类 BlockchainController ··············· 258
　9.7.6 创建文件 index.html ··············· 259
　9.7.7 修改配置文件 application.properties ··············· 259
　9.7.8 运行程序 ··············· 260
习题 9 ··············· 261

第 10 章　基于区块链的简易系统开发 ··············· 262

10.1 基于区块链的简易聊天室开发 ··············· 262
　10.1.1 操作界面 ··············· 262
　10.1.2 项目的主要文件构成 ··············· 262
　10.1.3 创建项目并添加依赖 ··············· 265

        10.1.4 创建类 Block …… 265
        10.1.5 创建类 Agent …… 267
        10.1.6 创建类 AgentServerThread …… 270
        10.1.7 创建类 AgentManager …… 272
        10.1.8 创建类 Message …… 273
        10.1.9 创建类 ChatController …… 274
        10.1.10 创建文件 index.html …… 275
        10.1.11 创建文件 display.js …… 276
        10.1.12 创建文件 restClient.js …… 278
        10.1.13 创建文件 main.css …… 279
        10.1.14 运行程序 …… 281
    10.2 基于区块链的简易证书系统开发 …… 281
        10.2.1 创建项目并添加依赖 …… 281
        10.2.2 创建类 Block …… 282
        10.2.3 创建类 Certificate …… 283
        10.2.4 创建接口 CertificateRepository …… 284
        10.2.5 创建类 CertificateService …… 284
        10.2.6 创建类 ByteUtils …… 285
        10.2.7 创建类 ProofOfWork …… 285
        10.2.8 创建类 PowResult …… 287
        10.2.9 创建类 SearchCertificateController …… 287
        10.2.10 创建文件 add_certificate.html …… 288
        10.2.11 创建文件 searchcer.html …… 289
        10.2.12 修改配置文件 application.properties …… 289
        10.2.13 运行程序 …… 290
    10.3 基于区块链的简易投票系统开发 …… 291
        10.3.1 创建项目并添加依赖 …… 291
        10.3.2 创建类 ElectionBlock …… 291
        10.3.3 创建类 Elections …… 292
        10.3.4 创建类 Voters …… 293
        10.3.5 创建类 Votes …… 293
        10.3.6 创建类 Candidates …… 294
        10.3.7 创建实体类访问数据库接口 …… 295
        10.3.8 创建类 HomeController …… 295
        10.3.9 创建类 CreatElectionController …… 296
        10.3.10 创建类 VoteController …… 298
        10.3.11 创建文件 index.html …… 301
        10.3.12 创建文件 add_election.html …… 302
        10.3.13 创建文件 add_voters.html …… 302

　　　　10.3.14　创建文件 add_vote.html …………………………………………… 303
　　　　10.3.15　创建文件 add_candidates.html ………………………………… 303
　　　　10.3.16　创建文件 view_votes.html ……………………………………… 304
　　　　10.3.17　修改配置文件 application.properties ………………………… 304
　　　　10.3.18　运行程序 …………………………………………………………… 304
　　习题 10 ……………………………………………………………………………………… 307
附录 A　Electrum 钱包的安装和配置 ………………………………………………… 308
附录 B　网页版 Coin.Space 钱包的创建 …………………………………………… 318
附录 C　JDK 的安装和配置 …………………………………………………………… 320
附录 D　IDEA 创建 Maven 多模块项目 ……………………………………………… 322
附录 E　ZooKeeper 的安装和配置 …………………………………………………… 326
附录 F　ActiveMQ 的下载与启动 …………………………………………………… 328
附录 G　RabbitMQ 的安装与配置 …………………………………………………… 330
附录 H　CouchDB 的安装与配置 …………………………………………………… 332
参考文献 ……………………………………………………………………………………… 334

第1章

区块链基础

本章包括区块链简介、区块链的发展、区块链的应用领域、区块链应用开发的相关技术、比特币钱包的安装与使用等内容。

1.1 区块链简介

本节介绍区块链的含义、特点和区块链的分类。

1.1.1 区块链的含义

区块链(Blockchain 或 Block Chain)是一种分布式的数据库系统,它采用点对点(Peer-to-Peer,P2P)协议存储、处理、传递、共享和同步带时序的数据。区块链可以是多中心、弱中心、分布式等架构形式。

在分布式系统中,一些分散的元素(如计算机、网络)组合在一起展现给用户一个系统、统一、完整的"错觉"。分布式系统拥有一些通用的物理和逻辑资源,可以动态分配任务;它们之间能够互相协调、互传消息来完成一个共同的任务。分布式系统需要一种方式(称为协议)来协调、同步分布式系统中的要素、资源,并且此协调过程对于用户是"透明"的(即用户不需要知道过程细节)。P2P 协议就是一种重要的分布式系统协议,也是区块链应用中所采用的协议。为了更好地互相协调完成任务,区块链中的数据是带有时序(加盖了时间戳)的数据。

为了保证数据的有效性,需要确保数据的安全性。区块链的应用基础是数据库(也称为账本),账本(Ledger)是"原始凭证"的隐喻。由于分布式系统中每个节点存储了账本,保证了全体数据(也称为总账)的真实性。每个节点处理的一条完整记录可以称为一个区块。由于节点之间的差异,使得区块结果是有差异的。正是由于这个原因,IOTA、Byteball、HashGraph 和 Intervalue 协议都提出了所谓的"无区块(Blockless)"的概念。

狭义的区块链技术是一种将时序数据组合成特定数据结构,并以密码学方式保证共享总账的不可篡改、不可伪造、弱中心化(Decentralized,也常译为去中心化),能够安全地存储、能在系统内验证的数据。

广义的区块链技术则是指一种全新的弱中心化的分布式系统,它利用加密技术来验证和存储数据,利用分布式共识算法来新增和更新数据,利用运行在区块链上的代码(即智能合约)来保证业务逻辑的自动执行。

1.1.2　区块链的特点

(1) 难以篡改。以比特币为例,除非掌握了51%以上的全网算力,否则不可能篡改区块链上的数据。目前比特币有数千种,遍布世界各地,且在一定程度上实现了比特币系统的不间断连续运行。如果交易数据已经被保存在足够多的区块之中,那么就可以丢弃该交易之前的数据。需要时,每个节点可以随时获取全部内容(即"同步"区块链中全部数据)。但是,区块链间会存在临时不一致的"分叉"现象和由于只有部分内容而导致的一些节点中数据之间的暂时"不同步"。

(2) 自由开放。任何人都可以参与到数据的创建、处理和传递等工作中来。从技术演进的角度来看,这是一个重大的进步。而且,对于开发人员来说,开源也是区块链应用的一大特点,这也是全局数据难以篡改这一特点的保障。

(3) 数据高度可信任。全局数据难以篡改的特点带来了数据的高可信性,因此基于这些可信数据可以进行多方面的应用和交易。

(4) 容错性强。区块链应用通过共识机制保持各个节点之间数据的高度一致,系统中存在多个数据副本;如果某个节点遇到网络问题、硬件故障、软件错误或者被黑客控制,均不会影响到整个系统或其他参与节点。问题节点在排除故障并完成数据同步之后,还可以随时再加入系统中继续工作。由于整个系统的正常运转不依赖于个别节点,所以每个节点可以有选择地下线,进行系统例行维护;同时还能保证整个系统7×24小时不间断地工作。

(5) 数据公开。比特币整个账本是公开的,但有些人或机构不愿意自己的资金交易(尤其是大额交易)被全网看到;隐私保护成为区块链技术的一个研究热点。一些解决方案已经出现,例如零币。

1.1.3　区块链的分类

(1) 按照开放程度,区块链可以分为公有链、联盟链和私有链。公有链中任何人都可以进行区块链数据的维护和读取,不受任何机构控制,系统最为开放。联盟链仅限联盟成员参与,是需要注册、许可才能访问的区块链。联盟规模可以是国与国之间,也可以是不同的组织之间。联盟链往往采用指定节点的记账方式,且进行记账的节点数量相对较少。私有链仅限于组织、国家机构内部或者单独个体使用,系统最为封闭。三类区块链的技术可以完全相同,只是开放程度、访问对象和访问权限存在差异。

(2) 按照应用范围,区块链可以分为基础链和行业链。所谓基础链就是提供有各类底层通用协议和工具、接口,方便开发者在此基础上快速开发出各种应用的区块链,一般以公有链为主。所谓行业链是指某些行业特别定制的基础协议和工具。

(3)按照独立程度,区块链可以分为主链和侧链。主链可以理解为正式上线的、独立的区块链网络。侧链不是特指某个区块链,而是遵守侧链协议的所有区块链的统称。

1.2 区块链的发展

本节介绍区块链的发展简史和区块链的发展阶段。

1.2.1 区块链的发展简史

2008年11月,中本聪发表了著名的论文《比特币:点对点的电子现金系统》。

2009年1月,中本聪用设计的比特币软件挖掘出了创始区块,开启了比特币的时代。

2010年9月,第一个"矿场"Slush发明了多个节点合作挖矿的方式,成为比特币挖矿这个行业的开端。

2011年4月,Bitcoin(是一种特定比特币的名称,不是通用比特币的英文翻译)第一个有官方正式记载的版本发布。在此之前Bitcoin节点支持的最小单位是0.01比特币,该版本真正支持"聪"这个比特币单位。2011年之后的几年时间里,莱特币、Ripple、R3等比特币和区块链技术竞相出现。

2013年,Bitcoin发布了0.8版本,完善了Bitcoin节点本身的内部管理,优化了网络通信。此后,Bitcoin才真正支持全网的大规模交易,成为中本聪设想的比特币。在Bitcoin 0.8版中引入了一个大bug,导致该版本发布后不久Bitcoin就出现了硬分叉,Bitcoin不得不回退到旧的版本,这也造成了Bitcoin价格产生大幅下跌。另外,硬分叉也说明了区块链是可回溯的时序数据库,有能防止伪造的特点。2014年发布了0.9版本,2015—2019年每年发布两个版本(即从0.10版到0.19版),2020年发布0.20版,2021年发布了0.21版和22.0版。

2014年,Vitalik Buterin发明以太坊。2015年7月发布正式的以太坊网络,标志着以太坊区块链正式上线运行;月底,以太币开始在多家交易所交易。2016年6月,以太坊上的一个去中心化自治组织The DAO被黑客攻击,市值五千万美元的以太币被转移。同年7月20日,以太币进行硬分叉,所有的以太币(包括被黑客转移的)回归原处。

超级账本(特指Hyperledger)是Linux基金会于2015年发起的推进区块链数字技术和交易验证的开源项目,其目标是让成员共同合作、共建开放平台,满足来自多个行业各种用户的需要。

1.2.2 区块链的发展阶段

1. 区块链1.0时代:比特币

区块链技术是伴随着比特币产生的,因此早期的区块链应用以比特币为代表。支撑比特币体系的主要技术包括哈希函数、分布式账本、区块链、非对称加密、工作量证明等,这些技术构成了区块链的最初版本。2010年7月著名比特币交易所Mt.gox成立,这标志着比特币真正进入了市场。尽管如此,能够了解到比特币、参与比特币买卖的用户仍较少。

2. 区块链2.0时代:技术金融

由于分布式账本技术是完全共享、透明和弱中心化的,故非常适合金融行业的应用。可以在区块链上发行多项资产;可进行货币以外的数字资产转移,如股票、债券。Bitcoin的造富效应带动了其他虚拟货币以及各种区块链应用的大爆发,出现众多百倍、千倍甚至万倍增殖的区块链资产,引发了全球的疯狂追捧。这使得比特币和区块链真正进入了全球视野。区块链2.0时代的主要特点是利用智能合约功能实现实时交互。

3. 区块链3.0时代:应用延伸

区块链2.0时代彻底颠覆了传统货币和支付的概念,而区块链3.0时代区块链被用于政府、公共卫生与医疗、科学与教育、文化与艺术、慈善捐赠等领域。区块链技术涉及P2P文件共享、分布式计算、网络模型、匿名等技术。

区块链的价值转移和信用转移优势得益于"弱中心化"的架构。"弱中心化"架构使得系统中的参与方不需要"信任中心"就可以完成交易和协作,这是传统互联网最薄弱的一项。区块链3.0时代的主要特点是"跨链",区块链的发展方向从金融转向医疗、保险、物流、游戏、域名等各个领域。

4. 区块链4.0时代:完整体系

区块链4.0首先要解决效率低下、能耗高、隐私保护、监管难等实际面临的问题,而且在未来的发展上要更具融合性。伴随着区块链数据容量的不断增加,数据的存储和确认成为制约区块链发展的难题,硬件区块链成为区块链4.0的主流趋势。

区块链1.0阶段是以比特币为代表的比特币应用,其主要职能是支持分布式账本和比特币流通;区块链2.0阶段开始在比特币与智能合约结合的金融领域得到更广泛的应用;区块链3.0阶段开始在政府、健康、文化和艺术、知识产权保护、信息安全管理等领域应用;区块链4.0阶段将成为人类日常生活应用的基础设施,基于区块链技术的完整体系将被建立。

1.3 区块链的应用领域

1.3.1 信息数据领域

信息数据领域不涉及实物,是最有可能产生创新的地方。未来应用的具体表现形态目前还很难预测,但是,至少可以应用于文件数据存储、游戏、数字版权保护等方面。

1.3.2 金融与数字资产领域

各国的金融系统已经在尝试使用区块链技术,应用领域主要包括支付系统、交易结算系统、资产证券化、信贷系统、保险系统等。

1.3.3 供应链领域

供应链是区块链和实业相关并能推动产业升级、可以大规模应用的领域,通过智能合约可实现可编程的商业。具体应用领域包括物联网、产品溯源、防伪、物流跟踪、消费者服务、供应链金融等。

1.3.4 政务领域

区块链的透明性和不可篡改性适合于政务系统。具体应用领域包括证书认证、工商登记、婚姻登记、房产登记、医疗记录、司法活动等。

正是由于区块链应用领域越来越广泛,所以虽然一些国家对比特币的交易进行了监管,但是对区块链应用的研究和开发却越来越多。

1.4 区块链应用开发的相关技术

本节先介绍区块链应用的整体技术模型,再分类介绍相关技术。

1.4.1 技术模型

区块链五大平台要素包括:共享账本、共识机制、隐私和安全、智能合约、商业网络。通过分析,本书将区块链应用的技术模型描述成如表1-1所示的表格。

表1-1 区块链应用的技术模型

层次	元素		系统	关键问题
应用系统	形式	钱包化	联盟化	数字化
	目标	个人应用一站化	价值网一体化	四流统一化
智能合约	实现	应用终端完成交易	工作流实现区块链	自定义区块语义
	架构	云计算	分布式系统	数据仓库和大数据
	程序	自动化	集成化	均衡化
信任机制	共识	PoW、PoS等	见证、连续6个确认	全社会信用记录
	交互	授权、认证与鉴权	同步、一致	P2P、并发
	算法	安全	数据记录、处理等	实时、响应式
支持环境	网络	社交网、物联网、通信网	互联网+	物流网
	基础	人、财、物、场地、工具等	方法、技术、制度等	标准、流程

1.4.2 支持环境

区块链是一种分布式应用系统,分布式系统的一些基础技术是其支撑环境。例如,计算机系统、网络设备、互联网协议、物联网、移动通信网络等。

1.4.3 信任机制

为了更好地分析信任机制的相关技术,本书将其细分为算法类技术、交互类技术、共识类技术。

其中,算法类技术包括加密与解密、哈希(Hash)函数、认证、数字签名等算法的分析、设计和实现,高并发和高安全性的编程语言(如 Rust)与算法,数据的记录、存储和处理等算法,实时系统和实时数据库、响应式(Reactive)开发等技术。

交互类技术包括系统和数据的授权、认证与鉴权(如人脸识别)、同步和异步、一致性、泛化的 P2P 协议(不仅仅是 P2P 网络)等技术。

共识类技术主要包括工作量证明(Proof of Work,PoW)、权益证明(Proof of Stack,PoS)和授权股权证明(Delegated Proof of Stack,DPoS)等算法。除此之外,还可以引入见证机制。以网上购物为例,一笔交易至少有买家、购物平台、卖家、支付平台、银行(可能涉及多家银行)、物流服务提供者(可能涉及多个物流系统)、监管部门(如税务部门、中央银行等)参与见证;而进行记账的角色至少有购物平台、支付平台、银行(可能多家)、物流服务提供者、税务部门(开发票后收税)等参与者(或系统)。在见证机制的基础上,还有一条原则(连续 6 个确认)可以来保证可信性。所谓连续 6 个确认是指在当前区块之后又有 5 个区块被计算出来并连接到区块链上,每个都相当于对前面一个区块进行确认,区块链上每增加一个区块就增大了前面区块被篡改的难度。在此基础上,实现全社会信用记录与共享,增加伪造、篡改等非法操作的成本,可以有效地保证彼此的信任。见证机制、连续 6 个确认原则以及全社会信用记录和共享都有赖于共享平台(如购物平台、支付平台、征信平台等)的实现。

1.4.4 智能合约

智能合约是一种自动触发机制,类似于计算机程序中 if-then 语句。为了更好地分析区块链中智能合约相关的技术,本书将其细分为程序类技术、架构类技术和实现类技术。

其中,程序类技术包括程序自动化、智能化、集成化(包括系统整合和数据融合)、平台化、均衡化(要实现数据共享、信息公开与隐私保护之间的均衡)等技术。架构类技术包括云计算、分布式系统(如微服务架构)、数据库(如 NoSQL)、数据仓库与数据挖掘(如商业智能)、大数据等技术。实现类技术包括应用终端实现交易、工作流等技术。

1.4.5 应用系统

实现区块链应用系统的目标是要实现个人应用一站化、价值网(上下游、前中后端)一体化、四流(业务流、资金流、信息流、物流)的统一化。具体表现形式为用户终端"钱包化"、个性化、移动化,价值网络呈现联盟化、网络化、增值化、共享化,四流呈现数字化。在传统模式下,以资金流为主要抓手促进物流,业务流为资金流和物流服务,信息流是辅助手段。在区块链模式下,业务流、资金流和信息流融合,共同促进物流。物流也大部分实现了信息化,即以信息化为核心促进物流。

1.4.6 区块链技术的多样性

结合以上分析，可以看出区块链技术呈现出多样性的特点。其代表性技术包括：

（1）不断发展的区块链技术与应用，如区块链即服务（Blockchain as a Service，BaaS）、智能合约、区块链应用系统等。

（2）分布式系统相关技术，如微服务架构、服务网格、无服务器架构（Serverless）等。其中，无服务器架构与区块链中"弱中心化"思想相似。

（3）数据库相关技术，如大数据、NoSQL 数据库、实时数据库、数据仓库和数据挖掘、JSON（JavaScript Object Notation，JavaScript 对象表示法）、XML（eXtensible Markup Language，可扩展标记语言）、DOM（Document Object Model，文档对象模型）等。

（4）促进业务流、资金流、物流和信息流四流合并的相关技术，如工作流技术、社会软件等。

（5）促进自动化的相关技术，如人工智能、生物特征识别（如人脸识别等）、物联网、互联网+等技术。

（6）终端系统和技术，如移动应用程序、微信小程序与二维码、云计算、虚拟现实与增强现实等。

（7）新的编程语言和开发工具、技术，如 Rust、Reactive。

（8）基础技术（如跨平台、数据融合和一致性、实时与并发、安全、P2P 等）。

（9）容器相关技术，如 Docker、Kubernetes 等。

通过以上分析，可以发现区块链的多样性特点和广泛应用决定了区块链技术的复杂性。因此，有必要深入研究、学习区块链应用开发。结合区块链应用的技术现状（多种开发语言、多种币种、多种平台、多种有待开发的应用领域、多参与者等），需要在区块链中有效地实现技术的兼容、跨域和融合。

1.5 比特币钱包的安装与使用

为了后面更好地理解区块链应用的基本需求，更好地进行区块链应用开发，本节介绍 Electrum 和 Coin.Space 等钱包的安装和使用。

1.5.1 Electrum 钱包的功能

PC 版 Electrum 钱包的详细安装和配置步骤见附录 A.1。完成安装后登录钱包，可以接收 Electrum，如图 1-1 所示；可以发送 Electrum，如图 1-2 所示；还能显示 Electrum 交易历史信息，如图 1-3 所示。接收 Electrum 时，有收款地址、请求的金额等内容；发送 Electrum 时，有收款地址、支付金额。网页版 Electrum 钱包的功能与 PC 版 Electrum 钱包的主要功能相同。网页版 Electrum 钱包的详细安装和配置步骤见附录 A.2。

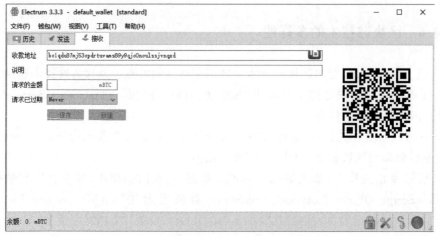

图 1-1 接收 Electrum 的功能界面

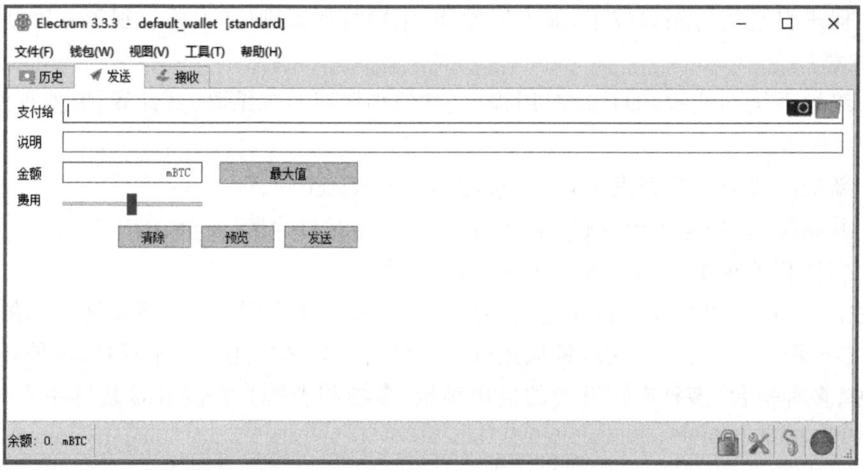

图 1-2 发送 Electrum 的功能界面

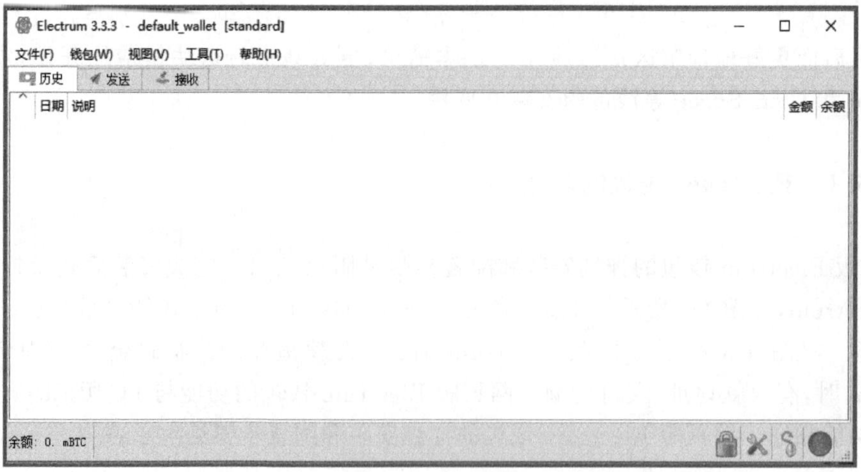

图 1-3 显示 Electrum 交易历史的功能界面

1.5.2 网页版 Coin.Space 钱包的功能

创建和配置 Coin.Space 钱包的详细步骤见附录 B。创建完成后登录钱包,能接收比特币,如图 1-4 所示;能发送比特币,如图 1-5 所示;还能显示比特币交易历史信息,如图 1-6 所示。

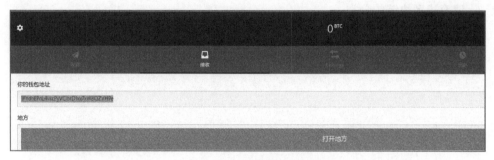

图 1-4 接收比特币的功能界面

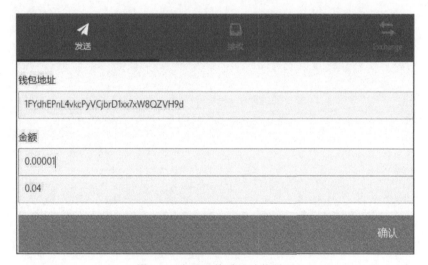

图 1-5 发送比特币的功能界面

图 1-6 显示比特币交易历史的功能界面

习题 1

简答题
1. 请简述区块链的含义。
2. 请简述区块链的特点。
3. 请简述区块链的分类。
4. 请简述区块链的应用领域。

第2章

Spring Boot开发基础

本书进行区块链应用开发的工具是 Spring Boot。本章先介绍 Spring Boot 基础；再介绍如何安装、配置 Spring Boot 的集成开发环境 IntelliJ IDEA（简称 IDEA），如何用 IDEA 创建项目与项目基本情况，Spring Boot 开发起步，如何实现 helloworld 项目。

2.1　Spring Boot 基础

本节先简要介绍 Spring Boot，再说明 Spring Boot 的主要特点。

2.1.1　Spring Boot 简介

Spring Boot 是伴随 Spring 4 而诞生的，它在继承 Spring 框架（简称 Spring）优点的基础上，简化了基于 Spring 的开发。Spring Boot 使得开发者可以更容易地创建基于 Spring 的"可以即时运行"的应用和服务。

Spring Boot 并不提供 Spring 的扩展功能，只是用于快速地开发基于 Spring 的应用程序。它并不是 Spring 的替代物，而是和 Spring 紧密结合来提高 Spring 应用开发者开发体验的工具。

从 2014 年 4 月发布 Spring Boot 1.0 到本书开始编写时，Spring Boot 的最新版本为 2.1.2 版。

2.1.2　Spring Boot 的主要特点

Spring Boot 使得创建独立（或产品级）的基于 Spring 应用变得更容易，大多数 Spring

Boot 应用只需要很少的 Spring 配置。开发 Spring Boot 的主要目的是为所有 Spring 开发者提供一个更迅速、可用的入门经验。Spring Boot 坚持"开箱即用",为许多项目提供一系列常用的非功能特性(例如安全、度量、健康检查、外部配置等)。

Spring Boot 的特点还包括:

(1) 约定大于配置。通过代码结构、注解的约定和命名规范等方式来减少配置,并采用更加简洁的配置方式来替代 XML 配置;减少冗余代码和过多的 XML 配置。

(2) 简化 Maven 配置。Maven 用于项目的构建,主要可以对依赖包进行管理,Maven 将项目中所依赖的 Jar 包信息放到 pom.xml 文件中< dependencies >和</ dependencies >节点之间。

(3) 内嵌有 Tomcat 等服务器,无须部署 War 文件。

(4) 使用注解使得编码变得更加简单。

(5) 对主流框架无配置集成,自动整合第三方框架,如 Struts。

2.2 配置 Spring Boot 开发环境

在进行 Spring Boot 开发之前,先要配置好开发环境。配置开发环境,需要先安装 JDK,然后选择安装一款合适的开发工具,本书以 IntelliJ IDEA 为开发工具。

2.2.1 安装和配置 JDK

安装和配置 JDK 的详细步骤见附录 C。完成安装后,打开 Windows 命令行程序,输入命令 java -version,如果见到如图 2-1 所示的版本信息就说明 JDK 安装成功了。

图 2-1 JDK 安装成功后显示的版本信息

2.2.2 安装 IDEA

可以从 IntelliJ IDEA 官方网站(https://www.jetbrains.com)下载免费的社区版或者试用的旗舰版 IDEA(在校学生和教师还可以免费使用旗舰版和教育版),然后进行安装。安装完成后打开 IDEA,将显示如图 2-2 所示的欢迎界面(本书使用的 IDEA 是旗舰版)。

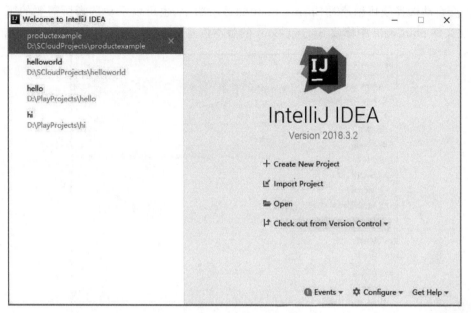

图 2-2　IDEA 启动后的欢迎界面

视频讲解

2.3　创建项目与项目基本情况

可以用不同的工具创建区块链应用项目,本节介绍利用 IDEA 创建项目的方法。

2.3.1　利用 IDEA 创建项目

先在如图 2-2 所示的欢迎界面中单击 Create New Project 链接,进入项目创建(New Project)界面,并选择 Spring Initializr 类型的项目,如图 2-3 所示。

接着,单击 Next 按钮跳转到项目元数据(Project Metadata)的设置界面,如图 2-4 所示。利用 IDEA 创建新项目时要根据情况填写(或设置)项目的元数据。

在图 2-4 所示的 Group 后面输入 com.bookcode,Artifact 后面输入 helloworld。在所创建项目的管理工具类型 Type 后面选择 Maven Project。由于目前 Maven 的参考资料比 Gradle 的参考资料更多且更容易获得,本书使用 Maven 进行项目管理。在开发语言 Language 后面选择 Java;打包方式 Packaging 后面选择 Jar;Java 的版本 Java Version 后面选择 8(也称为 1.8);所创建项目的版本 Version 后面保留自动生成的 0.0.1-SNAPSHOT;项目名称 Name 后面保留自动生成的 helloworld;项目描述 Description 后面可以修改为 Book Code for Blockchain with Spring Boot,将项目包名 Package 改为 com.bookcode。

填写完项目的元数据后,单击 Next 按钮就可以进入选择项目依赖(Dependencies)的界面,如图 2-5 所示。可以手动为所创建的项目(helloworld)选择项目依赖;也可以在创建项目时不选择任何依赖,而在文件 pom.xml 中添加所需要的依赖。选择完项目依赖的同时,

IDEA 会自动选择项目创建时 Spring Boot 的最新版本,也可以手动选择所需要的版本,还可以在文件 pom.xml 中修改 Spring Boot 的版本信息。

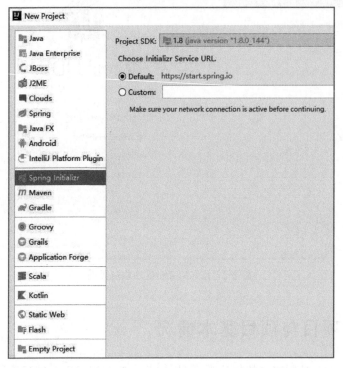

图 2-3　IDEA 中创建 New Project 时选择 Spring Initializr 类型项目的界面

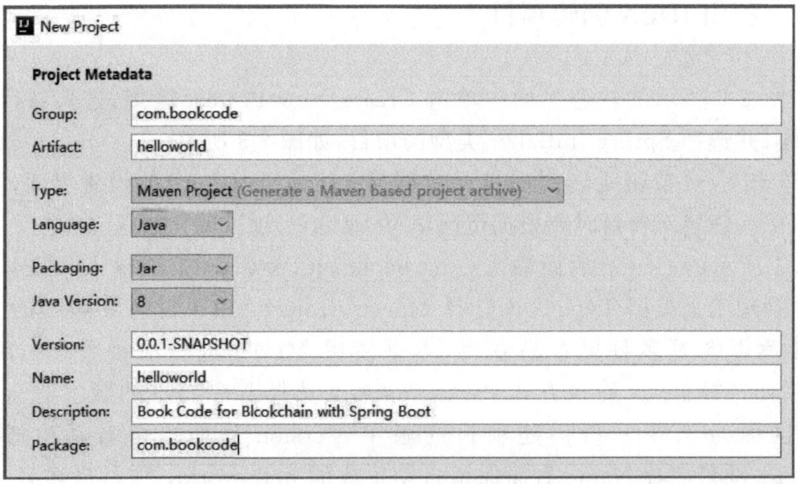

图 2-4　利用 IDEA 创建新项目时设置项目元数据(Project Metadata)的界面

单击 Next 按钮后,自动生成项目名称(Project name)和项目路径(Project location),可以保存默认生成的项目名称和项目路径不变;单击 Finish 按钮,就可以进入项目界面。由于所创建项目的管理类型为 Maven Project,所以文件 pom.xml 是项目中的关键文件,其代码如例 2-1 所示。其中,<parent>和</parent>之间的内容表示父依赖,是一般项目都要用到的基础内容,包含了项目中用到的 Spring Boot 的版本信息。<properties>和</properties>之间的内容表示了项目中所用到的 Java 版本信息。<dependencies>和</dependencies>之间的内容包含了项目所要用到的依赖信息,默认添加了对 spring-boot-starter(支持开箱即用)和 spring-boot-starter-test(支持测试)的依赖管理。关于 Maven 依赖的更多情况,将在后面的示例中逐步加以介绍。<build>和</build>之间的内容表示编译运行时要用到的相关插件。

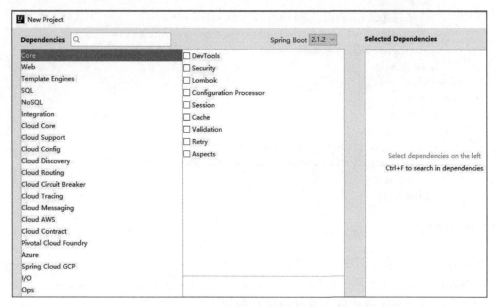

图 2-5 利用 IDEA 创建新项目时选择项目依赖(Dependencies)的界面

【例 2-1】 pom.xml 文件代码示例。

```
<?xml version = "1.0" encoding = "UTF-8"?>
<project xmlns = "http://maven.apache.org/POM/4.0.0"
xmlns:xsi = "http://www.w3.org/2001/XMLSchema-instance"
xsi:schemaLocation = "http://maven.apache.org/POM/4.0.0 http://maven.apache.org/xsd/maven-4.0.0.xsd">
    <modelVersion>4.0.0</modelVersion>
    <parent>
        <groupId>org.springframework.boot</groupId>
        <artifactId>spring-boot-starter-parent</artifactId>
<!-- 可以手动修改下面语句来修改 Spring Boot 的版本信息 -->
        <version>2.1.2.RELEASE</version>
        <relativePath/> <!-- lookup parent from repository -->
    </parent>
    <groupId>com.bookcode</groupId>
    <artifactId>springbootblockchain-helloworld</artifactId>
    <version>0.0.1-SNAPSHOT</version>
    <name>springbootblockchain-helloworld</name>
    <description>Book Code for Blockchain with Spring Boot</description>
    <properties>
```

```xml
        <java.version>1.8</java.version>
    </properties>
    <dependencies>
        <dependency>
            <groupId>org.springframework.boot</groupId>
            <artifactId>spring-boot-starter</artifactId>
        </dependency>
        <dependency>
            <groupId>org.springframework.boot</groupId>
            <artifactId>spring-boot-starter-test</artifactId>
            <scope>test</scope>
        </dependency>
    </dependencies>
    <build>
        <plugins>
            <plugin>
                <groupId>org.springframework.boot</groupId>
                <artifactId>spring-boot-maven-plugin</artifactId>
            </plugin>
        </plugins>
    </build>
</project>
```

至此，完成了项目的创建工作，在此基础上就可以进行 Spring Boot 开发了。为了内容的简洁，本书在后面章节的示例和案例中将不再介绍项目的创建过程。假如不太清楚项目创建过程，请先熟悉本节内容。

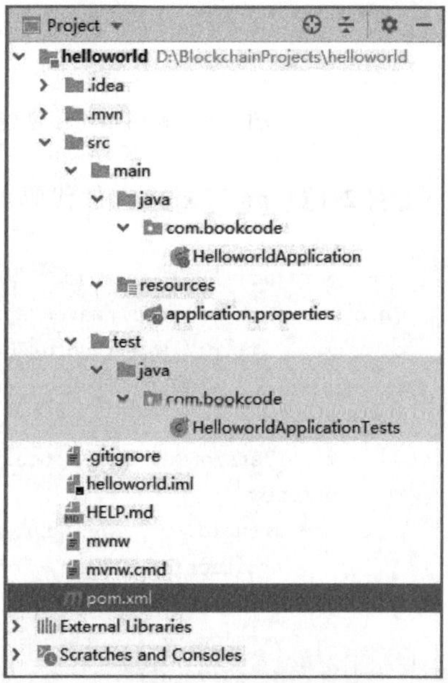

2.3.2 项目的基本构成

利用 IDEA 创建完项目之后，项目中目录和文件的构成情况如图 2-6 所示。项目中关键的目录、文件可以分为三大部分。其中，src/main/java 目录下包括主程序入口类 HelloworldApplication，可以通过该类来运行程序；开发时需要在此目录下添加所需的接口、类等文件。src/main/resources 是资源目录，该目录用来存放项目的资源，例如配置文件 application.properties。src/test 是单元测试目录，自动生成的测试文件 HelloworldApplicationTests 位于该目录下，用该测试文件可以测试程序。另外，pom.xml 文件是项目管理(特别是管理项目依赖)的重要文件。

图 2-6　利用 IDEA 创建项目后项目中目录和文件的构成情况

2.4 Spring Boot 开发起步

2.4.1 软件生命周期

为了应对软件危机,产生了软件工程学。软件工程是指导软件开发和维护的一门工程学科。采用工程的概念、原理、技术和方法来开发与维护软件,把有效的管理技术和适用的技术方法结合起来,经济地开发出高质量的软件并有效地维护它,这就是软件工程。

按照软件工程的理论,软件生命周期由软件定义、软件开发和运行维护3个时期组成,每个时期又进一步划分成若干阶段。

软件定义时期的任务包括:确定软件开发项目必须完成的总目标;确定项目的可行性;导出实现项目目标应该采用的策略及系统必须完成的功能;估算完成该项目所需要的资源和成本,并且制定项目进度表。软件定义时期通常可进一步划分成问题定义、可行性研究和需求分析3个阶段。

软件开发时期要设计和实现在前一个时期定义的软件,它通常由总体设计、详细设计、编码和单元测试、综合测试4个阶段组成。

运行维护时期的主要任务是使软件持久地满足用户的需要。通常维护活动可分为改正性维护、适应性维护、完善性维护和预防性维护4类。

由于本书主要探讨如何用Spring Boot开发区块链应用,所以本书的示例和案例主要说明的是如何用Spring Boot进行区块链应用的编码实现(简称为Spring Boot区块链应用开发)。

2.4.2 Spring Boot 开发步骤

Spring Boot区块链应用开发的步骤如下。

Step 1:打开开发工具;

Step 2:创建项目;

Step 3:判断是否需要添加依赖,如果不需要则跳过此步骤;

Step 4:创建类、接口(按照实体类、数据访问的接口和类、业务功能的接口和类、控制器类等顺序);

Step 5:判断是否需要创建视图文件和CSS等文件,如果不需要则跳过此步骤;

Step 6:判断是否需要创建、修改配置文件,如果不需要则跳过此步骤;

Step 7:判断是否需要图片、语音、视频等文件,如果不需要则跳过此步骤;

Step 8:判断是否需要下载辅助文件、包和安装工具(如数据库MySQL),如果不需要跳过此步骤。

由于本书中用到的工具安装过程比较简单,本书对工具安装的相关步骤介绍比较少。值得注意的是,Step 3～Step 8这6个步骤之间的顺序可以互换。

完成了开发之后,就可以运行程序了。

在区块链开发中可以按照不同应用层次(或方面)对要实现的功能进行细分后进行渐增式编程,如分为区块与挖矿、区块链、交易、钱包等功能后进行渐增式编程。细分之后,不仅易于开发且易于调试、测试。每一个功能开发步骤都相同,整个区块链开发过程是一个螺旋式上升的过程。

正是考虑到这一点,本书按照不同项目的重要程度(由基础到复杂)、开发实现难度(先易后难)的顺序组织各章内容,按照每个项目实现步骤的先后次序组织每章中的各节内容。

2.5 helloworld 项目的实现

视频讲解

本节介绍 helloworld 项目的实现。可以在 2.3.1 节创建的项目 helloworld 的基础上进行开发,也可以按照 2.3.1 节的步骤创建项目后进行开发。

2.5.1 添加依赖

确保在文件 pom.xml 的<dependencies>和</dependencies>之间添加了 Web、Gson 依赖,代码如例 2-2 所示。

【例 2-2】 添加依赖的代码示例。

```xml
<dependency>
        <groupId>org.springframework.boot</groupId>
        <artifactId>spring-boot-starter-web</artifactId>
</dependency>
<dependency>
        <groupId>com.google.code.gson</groupId>
        <artifactId>gson</artifactId>
        <version>2.8.5</version>
</dependency>
```

2.5.2 创建类 Block

在包 com.bookcode 中创建 entity 子包,在包 com.bookcode.entity 中创建类 Block,并修改类 Block 的代码,代码如例 2-3 所示。一般说来,创建类后都需要修改类的代码。故本书将创建类、修改类代码的过程简称为创建类,对应的类代码也默认为修改后的代码,后面章节也遵守此约定。

【例 2-3】 修改后的类 Block 代码示例。

```java
package com.bookcode.entity;
import com.bookcode.SecurityUtils;
import java.util.Date;
public class Block {
    public String hash;
    public String previousHash;
    private String data;
    private long timeStamp;
    private int nonce;
```

```
    public Block(String data, String previousHash) {
        this.data = data;
        this.previousHash = previousHash;
        this.timeStamp = new Date().getTime();
        this.hash = calculateHash();
    }
    public String calculateHash() {
         return SecurityUtils.encript(Integer.toString(nonce) + previousHash + Long.toString(timeStamp) + data);
    }
    public void mineBlock(int difficulty) {
        String target = new String(new char[difficulty]).replace('\0', '0');
        while (!hash.substring(0, difficulty).equals(target)) {
            nonce++;
            hash = calculateHash();
        }
        System.out.println("挖到的区块Hash代码是:" + hash + "!");
    }
}
```

2.5.3 创建类 SecurityUtils

在包 com.bookcode 中创建 utils 子包,并在包 com.bookcode.utils 中创建类 SecurityUtils(即包含创建类和修改类代码),代码如例 2-4 所示(默认为修改后的类代码)。

【例 2-4】 类 SecurityUtils 的代码示例。

```
package com.bookcode.utils;
import java.security.MessageDigest;
public class SecurityUtils {
    public static String encript(String input) {
        try {
            MessageDigest digest = MessageDigest.getInstance("SHA-256");
            byte[] hash = digest.digest(input.getBytes("UTF-8"));
            StringBuffer hexString = new StringBuffer();
            for (int i = 0; i < hash.length; i++) {
                String hex = Integer.toHexString(0xff & hash[i]);
                if (hex.length() == 1)
                    hexString.append('0');
                hexString.append(hex);
            }
            return hexString.toString();
        } catch (Exception e) {
            throw new RuntimeException(e);
        }
    }
}
```

2.5.4 创建类 HelloController

在包 com.bookcode 中创建 controller 子包，在包 com.bookcode.controller 中创建类 HelloController，代码如例 2-5 所示。

【例 2-5】 类 HelloController 的代码示例。

```java
package com.bookcode.controller;
import com.bookcode.springbootblockchainhelloworld.entity.Block;
import com.google.gson.GsonBuilder;
import org.springframework.web.bind.annotation.GetMapping;
import org.springframework.web.bind.annotation.RestController;
import java.util.ArrayList;
import java.util.List;
@RestController
public class HelloController {
    public static List<Block> blockchain = new ArrayList<Block>();
    public static int difficulty = 5;
    @GetMapping("/blockchain")
    public String blockchain() {
        blockchain.add(new Block("First block", "0"));
        System.out.println("挖第 1 个区块...");
        blockchain.get(0).mineBlock(difficulty);
        blockchain.add(new Block("Second block", blockchain.get(blockchain.size() - 1).hash));
        System.out.println("挖第 2 个区块...");
        blockchain.get(1).mineBlock(difficulty);
        System.out.println("区块链有效否: " + isValidChain());
        String blockchainJson = new GsonBuilder().setPrettyPrinting().create().toJson(blockchain);
        System.out.println("\n块链: ");
        System.out.println(blockchainJson);
        return "这是区块链.";
    }
    public static boolean isValidChain() {
        Block currentBlock;
        Block previousBlock;
        String targetHash = new String(new char[difficulty]).replace('\0', '0');
        for (int i = 1; i < blockchain.size(); i++) {
            currentBlock = blockchain.get(i);
            previousBlock = blockchain.get(i - 1);
            if (!currentBlock.hash.equals(currentBlock.calculateHash())) {
                System.out.println("Current Hashes not equal");
                return false;
            }
            if (!previousBlock.hash.equals(currentBlock.previousHash)) {
```

```java
                System.out.println("Previous Hashes not equal");
                return false;
            }
            if (!currentBlock.hash.substring(0, difficulty).equals(targetHash)) {
                System.out.println("This block hasn't been mined");
                return false;
            }
        }
        return true;
    }
}
```

2.5.5 运行程序

运行程序后，在浏览器中输入 localhost:8080/blockchain，浏览器中的输出结果如图 2-7 所示。控制台的主要输出结果如图 2-8 所示。

图 2-7　在浏览器中输入 localhost:8080/blockchain 后的结果

```
挖第 1 个区块...
挖到的区块 Hash 代码是：
0000026a71803dda86e111d5f98132413df75b0922180699de05f2140edbc1a5!
挖第 2 个区块...
挖到的区块 Hash 代码是：
000009a2ea245e5e23c62adfda1e569a44f819b2a48e5182cbe45e32bdae57ce!
区块链有效否: true
块链:
[
    {
        "hash":
"0000026a71803dda86e111d5f98132413df75b0922180699de05f2140edbc1a5",
        "previousHash": "0",
        "data": "First block",
        "timeStamp": 1551597071569,
        "nonce": 316974
    },
    {
        "hash":
"000009a2ea245e5e23c62adfda1e569a44f819b2a48e5182cbe45e32bdae57ce",
        "previousHash":
"0000026a71803dda86e111d5f98132413df75b0922180699de05f2140edbc1a5",
        "data": "Second block",
        "timeStamp": 1551597072939,
        "nonce": 388381
    }
]
```

图 2-8　在浏览器中输入 localhost:8080/blockchain 后控制台的主要输出结果

习题 2

一、简答题

1. 简述 Spring Boot 的主要特点。
2. 简述 Spring Boot 的开发步骤。

二、实验题

1. 安装 Spring Boot 的集成开发环境 IDEA。
2. 用 Spring Boot 实现 helloworld 项目的开发。

第3章 区块链应用的基础开发

本章介绍两个钱包之间的交互、挖矿、创建钱包和利用钱包进行交易、Merkle 树、Gossip 协议、IOTA(Internet of Things Application)等功能的简单模拟实现。

3.1 模拟两个钱包之间的交互

3.1.1 创建项目并添加依赖

视频讲解

参考 2.3.1 节的介绍，利用 IDEA 创建项目 basicexample。创建项目时，要添加 Web 依赖。假如创建项目时没有添加 Web 依赖，也可以在文件 pom.xml 的< dependencies >和</dependencies >之间直接添加 Web 依赖，代码如例 3-1 所示。

【例 3-1】 在文件 pom.xml 中添加依赖的代码示例。

```
< dependency >
        < groupId > org.springframework.boot </groupId >
        < artifactId > spring - boot - starter - web </artifactId >
</dependency >
```

3.1.2 创建接口 MainService

在包 com.bookcode 中创建 service 子包，并在包 com.bookcode.service 中创建接口 MainService 并修改接口代码，修改后的代码如例 3-2 所示。一般说来，创建接口后都需要修改接口的代码。故本书将创建接口并修改代码的过程简称为创建接口，对应的接口代码也默认为修改后的代码。后面的章节也遵守此约定。

【例 3-2】 修改后的接口 MainService 代码示例。

```
package com.bookcode.service;
public interface MainService {
    public String walletTo();
    public String block(boolean to);
}
```

3.1.3 创建类 MainServiceImpl

在包 com.bookcode.service 中创建 impl 子包,并在包 com.bookcode.service.impl 中创建类 MainServiceImpl,代码如例 3-3 所示。

【例 3-3】 类 MainServiceImpl 的代码示例。

```
package com.bookcode.service.impl;
import com.bookcode.service.MainService;
import org.slf4j.Logger;
import org.slf4j.LoggerFactory;
import org.springframework.beans.factory.annotation.Value;
import org.springframework.stereotype.Service;
import org.springframework.web.client.RestClientException;
import org.springframework.web.client.RestTemplate;
@Service
public class MainServiceImpl implements MainService {
    private static final Logger logger = LoggerFactory.getLogger(MainServiceImpl.class);
    @Value("${name}") //从配置文件中取出值
    private String name;
    @Value("${server.port}")
    private String port;
    @Value("${reciverPort}")
    private String reciverPort;
    @Value("${reciverName}")
    private String reciverName;
    @Value("${coins}")
    private String coins;
    @Override
    public String walletTo() {
        if (reciverPort == null)
            return "Next apps port not provided to app name: " + name;
        return callNext();
    }
    @Override
    public String block(boolean to) {
        if (to)
            return MsgSend();
        else
            return MsgRecive();
    }
    private String MsgSend() {
        String message = "success";
```

```
        message = "给在" + reciverPort + "的" + reciverName + "发送" + coins + "元
BCTestCoins.";
        return message;
    }
    private String MsgRecive() {
        String message = "success";
        message = "接收" + coins + "元 BCTestCoins.";
        return message;
    }
    private String callNext() {
        String message = "Error happened in app name: " + name;
        try {
            RestTemplate restTemplate = new RestTemplate();
            String url = "http://localhost:" + reciverPort + "/block/to";
        int response = restTemplate.getForEntity(url, String.class).hashCode();
//调用简单 Hash 函数
            message = "HashCode :" + response;
        } catch (RestClientException e) {
            e.printStackTrace();
        }
        return message;
    }
}
```

3.1.4　创建类 MainController

在包 com.bookcode 中创建 controller 子包，并在包 com.bookcode.controller 中创建类 MainController，代码如例 3-4 所示。

【例 3-4】　类 MainController 的代码示例。

```
package com.bookcode.controller;
import com.bookcode.service.MainService;
import org.springframework.web.bind.annotation.GetMapping;
import org.springframework.web.bind.annotation.RestController;
import javax.annotation.Resource;
@RestController
public class MainController {
    @Resource
    private MainService service;
    @GetMapping("/walletto")
    public String walletto() {
        return service.walletTo();
    }
    @GetMapping("/block/to")
    public String blockto() {
        return service.block(true);
    }
```

```
    @GetMapping("/block/receive")
    public String blockreceive() {
        return service.block(false);
    }
}
```

3.1.5 修改配置文件 application.properties

修改配置文件 application.properties，修改后的代码如例 3-5 所示。

【例 3-5】 修改后的配置文件 application.properties 代码示例。

```
server.port = 9000
name = 胡亥
reciverPort = 7000
reciverName = 刘邦
logging.level.org.springframework = INFO
coins = 4
BCTestCoins1ToETC = 6
tradeOrderNumber = 2019020600001
payForTrade = 2
difficulty = 5
```

3.1.6 运行程序

为了在同一台机器上模拟钱包的交互过程，用多个端口来运行程序。为了能并行运行多个端口，需要设置 IDEA 的运行方式（勾选 Allow running in parallel 复选框），如图 3-1 所示。

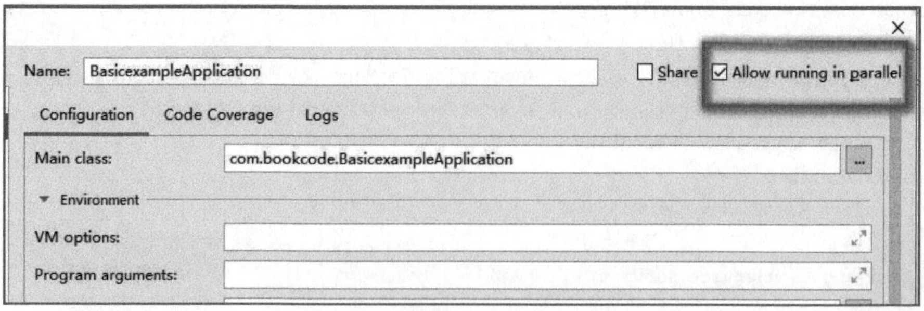

图 3-1 设置 IDEA 的运行方式

先运行已有程序，它的端口为 9000。在保持项目其他代码不变的情况下，再修改配置文件 application.properties，修改后的代码如例 3-6 所示。

【例 3-6】 修改后的配置文件 application.properties 代码示例。

```
server.port = 7000
name = 刘邦
reciverPort = 9000
reciverName = 胡亥
logging.level.org.springframework = INFO
coins = 4
BCTestCoins1ToETC = 6
tradeOrderNumber = 2019020600001
payForTrade = 2
difficulty = 5
```

修改完配置文件后再次运行程序,它的端口为7000。

在浏览器中输入 localhost：9000/walletto,结果如图 3-2 所示；在浏览器中输入 localhost：7000/walletto,结果如图 3-3 所示。接着,再在浏览器中输入 localhost：9000/walletto,结果如图 3-4 所示。对比图 3-2 和图 3-4 可以发现两次返回的哈希(Hash)值不一样,这里利用了 hash()函数的特点。hash()函数是一种从明文到密文的不可逆加密过程,无法根据结果计算出明文,从而保证数据的安全。而且,hash()函数结果值的长度越长越能保证数据的安全性。

图 3-2　在浏览器中输入 localhost：9000/walletto 的结果

图 3-3　在浏览器中输入 localhost：7000/walletto 的结果

在浏览器中输入 localhost：7000/block/to,结果如图 3-5 所示；在浏览器中输入 localhost：9000/block/to,结果如图 3-6 所示；在浏览器中输入 localhost：9000/block/recive,结果如图 3-7 所示。

图 3-4　再次在浏览器中输入 localhost：9000/walletto 的结果

图 3-5　在浏览器中输入 localhost：7000/block/to 的结果

图 3-6　在浏览器中输入 localhost：9000/block/to 的结果

图 3-7　在浏览器中输入 localhost：9000/block/recive 的结果

3.2 模拟挖矿

视频讲解

本节结合示例介绍挖矿的过程和结果的模拟实现。

3.2.1 添加依赖

继续在 3.1 节项目 basicexample 的基础上进行开发,在文件 pom.xml 的<dependencies>和</dependencies>之间直接添加 Lombok、Gson 依赖,代码如例 3-7 所示。

【例 3-7】 添加 Lombok、Gson 依赖的代码示例。

```xml
<dependency>
    <groupId>org.projectlombok</groupId>
    <artifactId>lombok</artifactId>
    <version>1.18.4</version>
</dependency>
<dependency>
    <groupId>com.google.code.gson</groupId>
    <artifactId>gson</artifactId>
    <version>2.8.5</version>
</dependency>
```

3.2.2 创建类 Block

在包 com.bookcode 中创建 entity 子包,并在包 com.bookcode.entity 中创建类 Block,代码如例 3-8 所示。

【例 3-8】 类 Block 的代码示例。

```java
package com.bookcode.entity;
import com.bookcode.util.StringUtil;
import lombok.Data;
import java.util.Date;
@Data //通过该注释省略代码中的 set()和 get()方法,例如类中省略了 getHash()和 SetHash()方法
public class Block {
    public String hash;
    public String previousHash;
    private String data;            //数据信息
    private long timeStamp;         //从 1970 年 1 月 1 日至今的毫秒数
    private int nonce;
    public Block(String data, String previousHash) {
        this.data = data;
```

```java
        this.previousHash = previousHash;
        this.timeStamp = new Date().getTime();
        this.hash = calculateHash(); //将此函数放在最后表明对所有内容进行加密
    }
    //计算新的矿块 Hash 值
    public String calculateHash() {
        String calculatedhash = StringUtil.applySha256(
                previousHash +
                        Long.toString(timeStamp) +
                        Integer.toString(nonce) +
                        data
        );
        return calculatedhash;
    }
    public void mineBlock(int difficulty) {
        String target = StringUtil.getDificultyString(difficulty);
        while(!hash.substring( 0, difficulty).equals(target)) {
            nonce ++;
            hash = calculateHash();
        }
        System.out.println("模拟挖矿,挖到矿块!!! Hash 值为: " + hash);
    }
}
```

3.2.3 创建类 MineService

在包 com.bookcode.service.impl 中创建类 MineService,代码如例 3-9 所示。

【例 3-9】 类 MineService 的代码示例。

```java
package com.bookcode.service.impl;
import com.bookcode.entity.Block;
import com.bookcode.util.StringUtil;
import org.springframework.beans.factory.annotation.Value;
import org.springframework.stereotype.Service;
import java.util.ArrayList;
@Service
public class MineService {
    private ArrayList<Block> blockchain = new ArrayList<Block>();
    @Value("${difficulty}")
    private String difficulty;
    public String mine() throws Exception {
        //增加区块到区块链中
        String bc = "\n 挖矿需要花费一些时间,请您耐心等待......\n 挖矿的结果为: \n";
        try {
```

```java
            System.out.println("尝试挖第 1 个区块……");
            addBlock(new Block("这是挖到的第 1 个区块", "0"));
            int diff = Integer.parseInt(difficulty);
            for (int i = 1; i < diff; i++) {
                int j = i + 1;
                System.out.println("尝试挖第" + j + "个区块……");
                String blockinfo = "这是挖到的第" + j + "……个区块";
                addBlock(new Block(blockinfo, blockchain.get(blockchain.size() - 1).hash));
            }
            System.out.println("\n区块链合法、有效: " + isChainValid());
            String blockchainJson = StringUtil.getJson(blockchain);
            System.out.println("\n区块链是: ");
            System.out.println(blockchainJson);
            bc += blockchainJson;
        } catch (Exception e) {
            e.printStackTrace();
        }
        return bc;
    }
    private Boolean isChainValid() {
        int diff = Integer.parseInt(difficulty);
        Block currentBlock;
        Block previousBlock;
        String hashTarget = new String(new char[diff]).replace('\0', '0');
        //通过 Hash 值判断区块链是否合法
        for (int i = 1; i < blockchain.size(); i++) {
            currentBlock = blockchain.get(i);
            previousBlock = blockchain.get(i - 1);
            //比较注册过的块的 Hash 值和当前块的 Hash 值是否相等
            if (!currentBlock.hash.equals(currentBlock.calculateHash())) {
                System.out.println("Current Hashes not equal");
                return false;
            }
            if (!previousBlock.hash.equals(currentBlock.previousHash)) {
                System.out.println("Previous Hashes not equal");
                return false;
            }
            if (!currentBlock.hash.substring(0, diff).equals(hashTarget)) {
                System.out.println("This block hasn't been mined");
                return false;
            }
        }
        return true;
    }
    private void addBlock(Block hi_im_the_first_block) {
        int diff = Integer.parseInt(difficulty);
        hi_im_the_first_block.mineBlock(diff);
```

```
        blockchain.add(hi_im_the_first_block);
    }
}
```

3.2.4 创建类 MineController

在包 com.bookcode.controller 中创建类 MineController，代码如例 3-10 所示。

【例 3-10】 类 MineController 的代码示例。

```
package com.bookcode.controller;
import com.bookcode.service.impl.MineService;
import org.springframework.beans.factory.annotation.Autowired;
import org.springframework.web.bind.annotation.GetMapping;
import org.springframework.web.bind.annotation.RestController;
@RestController
public class MineController {
    @Autowired
    private MineService mineService;
    @GetMapping("/mine")
    public String mine() throws Exception{
        return mineService.mine();
    }
}
```

3.2.5 运行程序

运行程序后，控制台的输出结果如图 3-8 所示。运行程序后，在浏览器中输入 localhost:9000/mine，结果如图 3-9 所示。通过图 3-9 可以看出，模拟的区块包括本区块的 Hash 值、前一区块的 Hash 值、数据、时间戳等信息。

```
尝试挖第1个区块……
模拟挖矿，挖到矿块!!!    Hash值为：000003eb57337c5d814136f12cdf0e964ceddb1f68cd70fb0660745ec216e7f4
尝试挖第2个区块……
模拟挖矿，挖到矿块!!!    Hash值为：00000b1dbc35bc341399273bac9dd778758bb90a7ced57ebf0fefa04207324cd
尝试挖第3个区块……
模拟挖矿，挖到矿块!!!    Hash值为：00000157b6135ecb691acd267ad0c8f26b1a879739903ccc9c2c5c7d9013cbd7
尝试挖第4个区块……
模拟挖矿，挖到矿块!!!    Hash值为：00000f3f11a3a6ac438ffdcaee5d55cd5dca7335be06b8d34bb09987131ef1dd
尝试挖第5个区块……
模拟挖矿，挖到矿块!!!    Hash值为：0000058d0e14a9be3da465c417db85fba11df3bd6f84498f76d168220c1112d6
区块链合法、有效：true
```

图 3-8 控制台的输出结果

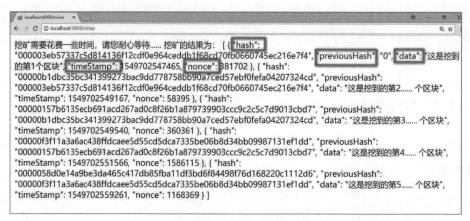

图 3-9　在浏览器中输入 localhost:9000/mine 后的结果

3.3　模拟创建钱包和利用钱包进行交易

视频讲解

　　本节结合示例介绍创建钱包和利用钱包进行交易。本节用到了 Web3j，Web3j 是一个高度模块化、响应式(Reactive)、类型安全的轻量级 Java 库，用于与 Ethereum 网络上的客户端(节点)集成。Web3j 使用 Ethereum 块链，而不需要为平台编写自己的集成代码的额外开销。

3.3.1　添加依赖

在 3.2 节项目 basicexample 的基础上，在文件 pom.xml 的 < dependencies > 和 </dependencies > 之间直接添加 Web3j 依赖，代码如例 3-11 所示。

【例 3-11】　添加 web 行依赖的代码示例。

```
<dependency>
        <groupId>org.web3j</groupId>
        <artifactId>web3j-spring-boot-starter</artifactId>
        <version>1.6.0</version>
</dependency>
```

3.3.2　创建类 StringUtil

在包 com.bookcode 中创建 util 子包，并在包 com.bookcode.util 中创建类 StringUtil，代码如例 3-12 所示。

【例 3-12】　类 StringUtil 的代码示例。

```
package com.bookcode.util;
import com.google.gson.GsonBuilder;
```

```java
import java.security.MessageDigest; //Java自带的数据加密类(包括MD5或SHA加密)
public class StringUtil {
    //对字符串进行SHA-256加密
    public static String applySha256(String input){
        try {
            MessageDigest digest = MessageDigest.getInstance("SHA-256");
            byte[] hash = digest.digest(input.getBytes("UTF-8"));
            StringBuffer hexString = new StringBuffer();
            for (int i = 0; i < hash.length; i++) {
                String hex = Integer.toHexString(0xff & hash[i]);
                if(hex.length() == 1) hexString.append('0');
                hexString.append(hex);
            }
            return hexString.toString();
        }
        catch(Exception e) {
            throw new RuntimeException(e);
        }
    }
    public static String getJson(Object o) {
        return new GsonBuilder().setPrettyPrinting().create().toJson(o);
    }
    public static String getDificultyString(int difficulty) {
        return new String(new char[difficulty]).replace('\0', '0');
    }
public static byte[] SHA2(byte[] b1, byte[] b2, byte[] b3, byte[] b4) {
        try {
            MessageDigest sha2 = MessageDigest.getInstance("SHA-256");
            byte[] b = new byte[b1.length + b2.length + b3.length + b4.length];
            System.arraycopy(b1, 0, b, 0, b1.length);
            System.arraycopy(b2, 0, b, b1.length, b2.length);
            System.arraycopy(b3, 0, b, b1.length + b2.length, b3.length);
            System.arraycopy(b4, 0, b, b1.length + b2.length + b3.length, b4.length);
            return sha2.digest(b);
        } catch (NoSuchAlgorithmException e) {
            e.printStackTrace();
            return null;
        }
    }
    public static byte[] SHA2(byte[] b) {
        try {
            MessageDigest sha2 = MessageDigest.getInstance("SHA-256");
            return sha2.digest(b);
        } catch (NoSuchAlgorithmException e) {
            e.printStackTrace();
            return null;
        }
    }
}
```

3.3.3 创建类 ContractService

在包 com.bookcode.service.impl 中创建类 ContractService,代码如例 3-13 所示。

【例 3-13】 类 ContractService 的代码示例。

```java
package com.bookcode.service.impl;
import org.slf4j.Logger;
import org.slf4j.LoggerFactory;
import org.springframework.beans.factory.annotation.Value;
import org.springframework.stereotype.Service;
import org.web3j.crypto.Credentials;
import org.web3j.crypto.WalletUtils;
import java.io.File;
import java.math.BigDecimal;
@Service
public class ContractService {
    private static final Logger log = LoggerFactory.getLogger(ContractService.class);
    @Value("${coins}")
    private String coins;
    @Value("${BCTestCoins1ToETC}")
    private String rate;
    @Value("${tradeOrderNumber}")
    private String orderNumber;
    @Value("${payForTrade}")
    private String pay;
    public String createWallet() throws Exception {
        File folder = new File("D:\\pictures"); //可以改成用户自己的目录地址
        String filePath = "d:/pictures";
        String fileName = "Can not create files.";
        try {
            String jsonfilename = WalletUtils.generateLightNewWalletFile("ws123", folder);
            int indexofhx = jsonfilename.lastIndexOf('-');
            int indexofd = jsonfilename.lastIndexOf('.');
            String address = jsonfilename.substring(indexofhx + 1, indexofd);
            fileName = "钱包地址为: 0x" + address;
//可以改成与用户自己的钱包地址相关的 JSON 文件,也可以采用源代码中的文件
String path = " d:/pictures/UTC--2019-02-06T10-11-28.47000000Z--28e27fc05ce14662990943872ae18deb231efaa8.json";
            Credentials ALICE = WalletUtils.loadCredentials("ws123", path);
            System.out.println("创建的钱包相关信息如下所示.");
            System.out.println("创建的第一个钱包地址为:" + ALICE.getAddress());
                                                                            //钱包地址
            System.out.println("密钥为:" + ALICE.getEcKeyPair().getPrivateKey()); //密钥
            System.out.println("公钥为:" + ALICE.getEcKeyPair().getPublicKey()); //公钥
            BigDecimal coinsBalance = new BigDecimal(coins);
            System.out.println("模拟向钱包充的 BCTestCoins 币: " + coinsBalance + "元.");
            BigDecimal changeRate = new BigDecimal(rate);
```

```
            BigDecimal balance = coinsBalance.multiply(changeRate);
            BigDecimal nbalance = balance.divide(new BigDecimal("100000"), 8, BigDecimal.ROUND_DOWN);
System.out.println("向钱包充的 BCTestCoins 币模拟兑换成以太币的数量为:" + nbalance + "元.");
        } catch (Exception e) {
            e.printStackTrace();
        }
        return fileName;
    }
    public String trade() throws Exception {
        String msg = "Trade Fail.";
        try {
             String path = "d:/pictures/UTC--2019-02-06T10-12-46.320000000Z--7ce63efbc015969b8d0a9355af615c0e85ed824a.json";
            Credentials ALICE = WalletUtils.loadCredentials("ws123", path);
            System.out.println("模拟的信息如下所示.");
            System.out.println("交易订单号为:" + orderNumber);
            System.out.println("收币人的钱包地址为:" + ALICE.getAddress());  //钱包地址
            System.out.println("收币人的公钥为:" + ALICE.getEcKeyPair().getPublicKey());
                                                                        //公钥
            BigDecimal payforT = new BigDecimal(pay);
            System.out.println("转账 BCTestCoins 币:" + payforT + "元.");
             String trademsg = "转给收币人(钱包地址为:" + ALICE.getAddress() + ")BCTestCoins 币" + payforT + "元.";
            System.out.println("进行 Hash 运算之前的交易转账信息:" + trademsg);
            int hc = trademsg.hashCode();
            msg = "交易转账信息进行 Hash 运算之后的结果:" + hc;
        } catch (Exception e) {
            e.printStackTrace();
        }
        return msg;
    }
}
```

3.3.4 创建类 ErcController

在包 com.bookcode.controller 中创建类 ErcController，代码如例 3-14 所示。

【例 3-14】 类 ErcController 的代码示例。

```
package com.bookcode.controller;
import com.bookcode.service.impl.ContractService;
import org.slf4j.Logger;
import org.slf4j.LoggerFactory;
import org.springframework.beans.factory.annotation.Autowired;
import org.springframework.web.bind.annotation.*;
@RestController
```

```java
public class ErcController {
    private static final Logger log = LoggerFactory.getLogger(ContractService.class);
    @Autowired
    ContractService contractService;
    @GetMapping("/create/wallet")
    public String createWallet() throws Exception {
        return contractService.createWallet();
    }
    @GetMapping("/trade")
    public String trade() throws Exception {
        return contractService.trade();
    }
}
```

3.3.5 运行程序

运行程序后,在浏览器中输入 localhost:9000/create/wallet,控制台的输出结果如图 3-10 所示;浏览器中的输出结果如图 3-11 所示。在浏览器中输出 localhost:9000/trade,控制台的输出结果如图 3-12 所示;浏览器中的输出结果如图 3-13 所示。

```
创建的钱包相关信息如下所示。
创建的第一个钱包地址为: 0x28e27fc05ce14662990943872ae18deb231efaa8
密钥为: 9585228245186026430831314845102132441503410361437274716548505970433978l049857
公钥为: 1296818168118500905501783413966601282872599762458067986756193422464821653815612...
模拟向钱包充的BCTestCoins币: 4元。
向钱包充的BCTestCoins币模拟兑换成以太币的数量为: 0.00024000元。
```

图 3-10 在浏览器中输入 localhost:9000/create/wallet 后控制台的输出结果

图 3-11 在浏览器中输入 localhost:9000/create/wallet 后浏览器中的输出结果

```
模拟的信息如下所示。
交易订单号为: 2019020600001
收币人的钱包地址为: 0x7ce63efbc015969b8d0a9355af615c0e85ed824a
收币人的公钥为: 50377703711249285103544635629318377361461910030284493176534441671500611625805289905505404974309054422 89
转账BCTestCoins币: 2 元。
进行Hash运算之前的交易转账信息: 转给收币人(钱包地址为: 0x7ce63efbc015969b8d0a9355af615c0e85ed824a) BCTestCoins币2元。
```

图 3-12 在浏览器中输入 localhost:9000/trade 后控制台的输出结果

图 3-13　在浏览器中输入 localhost:9000/trade 后浏览器中的输出结果

3.4　Merkle 树的实现

视频讲解

本节介绍 Merkle 树的实现。本节是在 3.3 节 basicexample 项目的基础上进行开发的，其中用到的辅助类 StringUtil 代码和 3.3 节相同。

3.4.1　创建类 Stakeholder

在包 com.bookcode.entity 中创建类 Stakeholder，代码如例 3-15 所示。

【例 3-15】　类 Stakeholder 的代码示例。

```java
package com.bookcode.entity;
import java.nio.charset.StandardCharsets;
//Merkle 树中有 Stakeholder
public class Stakeholder {
    private final String name;
    private final int coins;
    public Stakeholder(String name, int coins) {
        this.name = name;
        this.coins = coins;
    }
    public String getName() {
        return name;
    }
    public int getCoins() {
        return coins;
    }
    public byte[] toBytes() {
        return String.format("%s%d", name, coins).
                getBytes(StandardCharsets.UTF_8);
    }
    @Override
    public String toString() {
        return name;
    }
}
```

3.4.2 创建类 ProofEntry

在包 com.bookcode.entity 中创建类 ProofEntry,代码如例 3-16 所示。

【例 3-16】 类 ProofEntry 的代码示例。

```
package com.bookcode.entity;
import javax.xml.bind.DatatypeConverter;
public class ProofEntry {
    private final byte[] hash;
    private final int x1, x2;
    public ProofEntry(byte[] hash, int x1, int x2) {
        this.hash = hash;
        this.x1 = x1;
        this.x2 = x2;
    }
    public int getLeftBound() {
        return x1;
    }
    public int getRightBound() {
        return x2;
    }
    public byte[] getMerkleHash() {
        return hash;
    }
    @Override
    public String toString() {
        return String.format("(%s, %d, %d)",
                DatatypeConverter.printHexBinary(hash),
                x1,
                x2);
    }
}
```

3.4.3 创建类 Node

在包 com.bookcode.entity 中创建类 Node,代码如例 3-17 所示。

【例 3-17】 类 Node 的代码示例。

```
package com.bookcode.entity;
import com.bookcode.util.StringUtil;
public class Node {
    private final Node left, right;
    private final Stakeholder stakeholder;
    private final byte[] hash;
    private Node(Node left, Node right, Stakeholder stakeholder, byte[] hash) {
```

```java
        this.stakeholder = stakeholder;
        this.left = left;
        this.right = right;
        this.hash = hash;
    }
    public Node(Node left, Node right, byte[] hash) {
        this(left, right, null, hash);
    }
    public Node(Stakeholder stakeholder) {
        this(null, null, stakeholder, StringUtil.SHA2(stakeholder.toBytes()));
    }
    public boolean isLeaf() {
        return stakeholder != null;
    }
    public Stakeholder getStakeholder() {
        return stakeholder;
    }
    public Node getLeftNode() {
        return left;
    }
    public Node getRightNode() {
        return right;
    }
    public byte[] getMerkleHash() {
        return hash;
    }
    public int getCoins() {
        if (isLeaf()) {
            return stakeholder.getCoins();
        }
        return left.getCoins() + right.getCoins();
    }
}
```

3.4.4 创建类 FtsResult

在包 com.bookcode.entity 中创建类 FtsResult,代码如例 3-18 所示。

【例 3-18】 类 FtsResult 的代码示例。

```java
package com.bookcode.entity;
import java.util.List;
public class FtsResult {
    private final List<ProofEntry> merkleProof;
    private final Stakeholder stakeholder;
    public FtsResult(List<ProofEntry> merkleProof, Stakeholder stakeholder) {
        this.merkleProof = merkleProof;
        this.stakeholder = stakeholder;
```

```java
    }
    public Stakeholder getStakeholder() {
        return stakeholder;
    }
    public List<ProofEntry> getMerkleProof() {
        return merkleProof;
    }
    @Override
    public String toString() {
        final StringBuilder sb = new StringBuilder();
        sb.append("merkleProof {\n");
        for (ProofEntry proofEntry : merkleProof) {
            sb.append(String.format("    %s\n", proofEntry.toString()));
        }
        sb.append("}\n");
        sb.append("stakeholder {\n");
        sb.append(String.format("    %s\n", stakeholder.toString()));
        sb.append("}");
        return sb.toString();
    }
}
```

3.4.5 创建类 FtsService

在包 com.bookcode.service.impl 中创建类 FtsService,代码如例 3-19 所示。

【例 3-19】 类 FtsService 的代码示例。

```java
package com.bookcode.service.impl;
import com.bookcode.entity.FtsResult;
import com.bookcode.entity.Node;
import com.bookcode.entity.ProofEntry;
import com.bookcode.entity.Stakeholder;
import com.bookcode.util.StringUtil;
import org.springframework.stereotype.Service;
import javax.xml.bind.DatatypeConverter;
import java.nio.charset.StandardCharsets;
import java.util.ArrayList;
import java.util.Arrays;
import java.util.List;
import java.util.Random;
@Service
public class FtsService {
    public String mtree() {
        //创建 stakeholders
        List<Stakeholder> stakeholders = new ArrayList<>();
        for (int i = 0, c = 2; i < 3; i++, c = c % 2 == 0 ? c / 2 : c * 3 + 1) {
            final String name = String.format("Stakeholder %d", i);
```

```java
            final int coins = c;
            stakeholders.add(new Stakeholder(name, coins));
        }
        //创建 Merkle 树
        final Node[] tree = createMerkleTree(stakeholders);
        System.out.println("Doing follow-the-satoshi in the stake tree.");
        FtsResult ftsResult = ftsTree(tree, new Random(4));
        System.out.println(ftsResult);
        System.out.println("Verifying the result.");
        ftsVerify(new Random(42), tree[1].getMerkleHash(), ftsResult);
        return "Created Merkle Tree Success.";
    }
    public Node[] createMerkleTree(List<Stakeholder> stakeholders) {
        final Node[] tree = new Node[stakeholders.size() * 2];
        System.out.println(String.format("Creating Merkle tree with %d nodes.",
tree.length - 1));
        for (int i = 0; i < stakeholders.size(); i++) {
            tree[stakeholders.size() + i] = new Node(stakeholders.get(i));
        }
        for (int i = stakeholders.size() - 1; i > 0; i--) {
            final Node left = tree[i * 2];
            final Node right = tree[i * 2 + 1];
            final byte[] hash = StringUtil.SHA2(left.getMerkleHash(),
                    right.getMerkleHash(),
                    String.valueOf(left.getCoins()).getBytes(StandardCharsets.UTF_8),
                    String.valueOf(right.getCoins()).getBytes(StandardCharsets.UTF_8));
            tree[i] = new Node(left, right, hash);
        }
        for (int i = 1; i < tree.length; i++) {
System.out.println(String.format("Hash %d: %s", i, DatatypeConverter.printHexBinary(tree[i].
getMerkleHash())));
        }
        return tree;
    }
    public FtsResult ftsTree(Node[] tree, Random rng) {
        int i = 1;
        List<ProofEntry> merkleProof = new ArrayList<>();
        while (true) {
            if (tree[i].isLeaf()) {
                return new FtsResult(merkleProof, tree[i].getStakeholder());
            }
            final int x1 = tree[i].getLeftNode().getCoins();
            final int x2 = tree[i].getRightNode().getCoins();
         System.out.println(String.format("Left subtree %d coins / right subtree %d
coins.", x1, x2));
            final int r = rng.nextInt(x1 + x2) + 1;
            System.out.println(String.format("Picking coin number %d", r));
            if (r <= x1) {
                System.out.println("Choosing left subtree...");
                i *= 2;
```

```java
                merkleProof.add(new ProofEntry(tree[i + 1].getMerkleHash(), x1, x2));
            } else {
                System.out.println("Choosing right subtree...");
                i = i * 2 + 1;
                merkleProof.add(new ProofEntry(tree[i - 1].getMerkleHash(), x1, x2));
            }
        }
    }
    public boolean ftsVerify(Random rng, byte[] merkleRootHash, FtsResult ftsResult) {
        StringBuilder auditPath = new StringBuilder();
        System.out.print("Building audit path... ");
        for (ProofEntry proofEntry : ftsResult.getMerkleProof()) {
            final int x1 = proofEntry.getLeftBound();
            final int x2 = proofEntry.getRightBound();
            final int r = rng.nextInt(x1 + x2) + 1;
            if (r <= x1) {
                System.out.print("0 ");
                auditPath.append('0');
            } else {
                System.out.print("1 ");
                auditPath.append('1');
            }
        }
        System.out.println("OK");
        byte[] hx = StringUtil.SHA2(ftsResult.getStakeholder().toBytes());
        for (int i = ftsResult.getMerkleProof().size() - 1; i >= 0; i--) {
            final ProofEntry proofEntry = ftsResult.getMerkleProof().get(i);
final byte[] x1 = String.valueOf(proofEntry.getLeftBound()).getBytes(StandardCharsets.UTF_8);
final byte[] x2 = String.valueOf(proofEntry.getRightBound()).getBytes(StandardCharsets.UTF_8);
            final byte[] hy = proofEntry.getMerkleHash();
            if (auditPath.charAt(i) == '0') {
                hx = StringUtil.SHA2(hx, hy, x1, x2);
            } else {
                hx = StringUtil.SHA2(hy, hx, x1, x2);
            }
System.out.println(String.format("Next hash: %s", DatatypeConverter.printHexBinary(hx)));
        }
        boolean result = Arrays.equals(hx, merkleRootHash);
        System.out.println(result ? "Root hash matches!" : "Invalid Merkle proof.");
        return result;
    }
}
```

3.4.6 创建类 FtsController

在包 com.bookcode.controller 中创建类 FtsController,代码如例 3-20 所示。

【例 3-20】 类 FtsController 的代码示例。

```
package com.bookcode.controller;
import com.bookcode.service.impl.FtsService;
import org.springframework.beans.factory.annotation.Autowired;
import org.springframework.web.bind.annotation.GetMapping;
import org.springframework.web.bind.annotation.RestController;
@RestController
public class FtsController {
    @Autowired
    FtsService service;
    @GetMapping("/mtree")
    public String mtree() throws Exception {
        return service.mtree();
    }
}
```

3.4.7　运行程序

运行程序后，在浏览器中输入 localhost：9000/mtree，浏览器的输出结果如图 3-14 所示；控制台的输出结果如图 3-15 所示。

图 3-14　在浏览器中输入 localhost：9000/mtree 后浏览器的输出结果

```
Creating Merkle tree with 5 nodes.
Hash 1: 118C7EEA7A0AB327F5400DFFA1FEF04F67C3CD7F3E1D33F957AA697332D267BE
Hash 2: C4492F1A0EA8C3E13092B5BF18B5FC1256262C037451FF359E7D50F28E634617
Hash 3: 1CF87AA2BAEDD7F5D16EE7F67567063FDD7EF861862A1B6547E91E99140EB03B
Hash 4: E697BCE025BA1173069D4ED21D1A792BDDC57169BEC5BEEA4B1888CC43CF7395
Hash 5: 4F0C619782DEBF8B2F28DAFE4B4110854CEC0FC5D6238AD3DB8F30B9C4CEA970
Doing follow-the-satoshi in the stake tree.
Left subtree 5 coins / right subtree 2 coins.
Picking coin number 4
Choosing left subtree...
Left subtree 1 coins / right subtree 4 coins.
Picking coin number 3
Choosing right subtree...
merkleProof {
    (1CF87AA2BAEDD7F5D16EE7F67567063FDD7EF861862A1B6547E91E99140EB03B, 5, 2)
    (E697BCE025BA1173069D4ED21D1A792BDDC57169BEC5BEEA4B1888CC43CF7395, 1, 4)
}
stakeholder {
    Stakeholder 2
}
Verifying the result.
Building audit path... 0 1 OK
Next hash: C4492F1A0EA8C3E13092B5BF18B5FC1256262C037451FF359E7D50F28E634617
Next hash: 118C7EEA7A0AB327F5400DFFA1FEF04F67C3CD7F3E1D33F957AA697332D267BE
Root hash matches!
```

图 3-15　在浏览器中输入 localhost：9000/mtree 后控制台的输出结果

3.5 Gossip 协议的模拟实现

视频讲解

Gossip 协议是分布式系统中被广泛使用的协议,其主要用于实现分布式节点或者进程之间的信息交换。区块链中区块间可以通过 Gossip 等协议进行通信。本节基于已有的 Gossip 库模拟实现 Gossip 协议。

3.5.1 创建项目并添加依赖

用 IDEA 创建完项目 exgossip 之后,确保在文件 pom.xml 的 <dependencies> 和 </dependencies> 之间添加了 Web、Gossip、Lombok 依赖,代码如例 3-21 所示。

【例 3-21】 添加 Web、Gossip、Lombok 依赖的代码示例。

```xml
<dependency>
    <groupId>org.springframework.boot</groupId>
    <artifactId>spring-boot-starter-web</artifactId>
</dependency>
<dependency>
    <groupId>org.apache.gossip</groupId>
    <artifactId>gossip</artifactId>
    <version>0.1.2-incubating</version>
</dependency>
<dependency>
    <groupId>org.projectlombok</groupId>
    <artifactId>lombok</artifactId>
    <version>1.18.6</version>
</dependency>
```

3.5.2 创建类 Node

在包 com.bookcode 中创建 gprotocol 子包,并在包 com.bookcode.gprotocol 中创建类 Node,代码如例 3-22 所示。

【例 3-22】 类 Node 的代码示例。

```java
package com.bookcode.gprotocol;
import lombok.Data;
import java.net.URI;
@Data
public class Node {
    private final String id;
    private final String ipAddress;
    private final int port;
    public Node(String id, String ipAddress, int port) {
        this.id = id;
```

```
            this.ipAddress = ipAddress;
            this.port = port;
        }
        public URI getURI() {
            return URI.create("udp://localhost:" + String.valueOf(port));
        }
    }
```

3.5.3 创建类 StaticSeedFinder

在包 com.bookcode.gprotocol 中创建类 StaticSeedFinder，代码如例 3-23 所示。

【例 3-23】 类 StaticSeedFinder 的代码示例。

```
package com.bookcode.gprotocol;
import lombok.Data;
import java.util.HashSet;
import java.util.Set;
@Data
public class StaticSeedFinder {
    private final String ipAddress;
    private final int port;
    private final String id;
    public StaticSeedFinder(String id, String ipAddress, int port) {
        this.ipAddress = ipAddress;
        this.port = port;
        this.id = id;
    }
    public Set<Node> lookup() {
        Node seed = new Node(id, ipAddress, port);
        Set<Node> seeds = new HashSet<>();
        seeds.add(seed);
        return seeds;
    }
    @Override
    public String toString() {
        return "id = " + getId() + ", ipAddress = " + getIpAddress() + ", port = " + getPort();
    }
}
```

3.5.4 创建类 SeedFinderChain

在包 com.bookcode.gprotocol 中创建类 SeedFinderChain，代码如例 3-24 所示。

【例 3-24】 类 SeedFinderChain 的代码示例。

```java
package com.bookcode.gprotocol;
import org.springframework.beans.factory.annotation.Autowired;
import java.util.HashSet;
import java.util.List;
import java.util.Set;
public class SeedFinderChain {
    private List<StaticSeedFinder> staticSeedFinders;
    @Autowired
    public SeedFinderChain(List<StaticSeedFinder> seedFinders) {
        this.staticSeedFinders = seedFinders;
    }
    public Set<Node> lookup() {
        Set<Node> seeds = new HashSet<>();
        for (StaticSeedFinder seedFinder : staticSeedFinders) {
            seeds.addAll(seedFinder.lookup());
        }
        return seeds;
    }
    @Override
    public String toString() {
        String strResult = "[";
        for(StaticSeedFinder ssf: staticSeedFinders){
            strResult += "{" + ssf.toString() + " }";
        }
        return "节点包括: " + strResult + "].";
    }
}
```

3.5.5 创建类 GossipConnector

在包 com.bookcode.gprotocol 中创建类 GossipConnector，代码如例 3-25 所示。

【例 3-25】 类 GossipConnector 的代码示例。

```java
package com.bookcode.gprotocol;
import org.apache.gossip.GossipSettings;
public class GossipConnector {
    private final Node host;
    private final Node seed;
    public GossipConnector(Node host, Node seed) {
        this.host = host;
        this.seed = seed;
    }
    public void listen() {
        GossipSettings s = new GossipSettings();
        s.setGossipInterval(100);
```

```
            System.out.println("Registering host: " + host.toString());
            System.out.println("Registering seed: " + seed.toString());
    }
}
```

3.5.6 创建类 GossipController

在包 com.bookcode 中创建 controller 子包，并在包 com.bookcode.controller 中创建类 GossipController，代码如例 3-26 所示。

【例 3-26】 类 GossipController 的代码示例。

```
package com.bookcode.controller;
import com.bookcode.gprotocol.GossipConnector;
import com.bookcode.gprotocol.Node;
import com.bookcode.gprotocol.SeedFinderChain;
import com.bookcode.gprotocol.StaticSeedFinder;
import org.springframework.web.bind.annotation.GetMapping;
import org.springframework.web.bind.annotation.RestController;
import java.util.ArrayList;
import java.util.List;
import java.util.Set;
@RestController
public class GossipController {
    @GetMapping("/testGP")
    public String blockchain() {
        StaticSeedFinder staticSeedFinder = new StaticSeedFinder("1","131.1.2.1",8060);
        StaticSeedFinder staticSeedFinder1 = new StaticSeedFinder ("1","147.132.107.102",8000);
        List<StaticSeedFinder> seedFinders = new ArrayList<>();
        Set<Node> sns = staticSeedFinder.lookup();
        System.out.println(sns.toString());
        seedFinders.add(staticSeedFinder);
        seedFinders.add(staticSeedFinder1);
        List<StaticSeedFinder> ssf = seedFinders.subList(1,2);
        for(StaticSeedFinder s:ssf){
            System.out.println(s.toString());
        }
        SeedFinderChain seedFinderChain = new SeedFinderChain(ssf);
        System.out.println(seedFinderChain.toString());
        Node node = new Node("1","127.0.0.1",80);
        System.out.println("节点的 URI 为: " + node.getURI());
        Node node1 = new Node("1","131.1.8.1",8000);
        GossipConnector gossipConnector = new GossipConnector(node,node1);
        gossipConnector.listen();
        return "成功模拟 Gossip 协议.";
    }
}
```

3.5.7 运行程序

运行程序后,在浏览器中输入 localhost:8080/testGP,浏览器的输出结果如图 3-16 所示;控制台的输出结果如图 3-17 所示。

图 3-16 在浏览器中输入 localhost:8080/testGP 后浏览器的输出结果

```
[Node(id=1, ipAddress=131.1.2.1, port=8060)]
id=1, ipAddress=147.132.107.102, port=8000
节点包括:[{id=1, ipAddress=147.132.107.102, port=8000 }]。
节点的URI为: udp://localhost:80
Registering host: Node(id=1, ipAddress=127.0.0.1, port=80)
Registering seed: Node(id=1, ipAddress=131.1.8.1, port=8000)
```

图 3-17 在浏览器中输入 localhost:8080/testGP 后控制台的输出结果

3.6 模拟 IOTA 的应用

视频讲解

IOTA(Internet of Things Application)是一种新型的分布式网络应用,专注于解决机器和机器之间(Machine to Machine,M2M)的交易问题。它使用独特的方法来验证交易,使其成为物联网所需的大量数据交换的理想选择。从底层网络的角度来看,它的核心是基于 Gossip 协议的交易传播。本节利用已有的 Jota 库来模拟 IOTA 的应用。

3.6.1 创建项目并添加依赖

用 IDEA 创建完项目 exiota 之后,确保在文件 pom.xml 的<dependencies>和</dependencies>之间添加了 Web、Gson、Jota 依赖,代码如例 3-27 所示。

【例 3-27】 添加 Web、Gson、Jota 依赖的代码示例。

```
< dependency >
        < groupId > org.springframework.boot </groupId >
        < artifactId > spring - boot - starter - web </artifactId >
</dependency >
< dependency >
        < groupId > com.google.code.gson </groupId >
        < artifactId > gson </artifactId >
        < version > 2.8.5 </version >
</dependency >
```

```xml
<dependency>
        <groupId>org.iota</groupId>
        <artifactId>jota</artifactId>
        <version>1.0.0-beta3</version>
</dependency>
```

3.6.2 创建类 IotaController

在包 com.bookcode 中创建 controller 子包,并在包 com.bookcode.controller 中创建类 IotaController,代码如例 3-28 所示。

【例 3-28】 类 IotaController 的代码示例。

```java
package com.bookcode.controller;
import jota.IotaAPI;
import org.springframework.stereotype.Controller;
import org.springframework.web.bind.annotation.GetMapping;
import org.springframework.web.bind.annotation.PathVariable;
@Controller
public class IotaController {
    @GetMapping("/test/{protocol}/{host}/{port}")
    String test(@PathVariable String protocol, @PathVariable String host, @PathVariable String port) {
        IotaAPI api = new IotaAPI.Builder()
                .protocol(protocol)
                .host(host)
                .port(port).build();
        String strApiProtocol = api.getProtocol();
        System.out.println("协议:" + strApiProtocol);
        String strApiHost = api.getHost();
        System.out.println("主机:" + strApiHost);
        String strApiPort = api.getPort();
        System.out.println("端口:" + strApiPort);
        String strURL = strApiProtocol + "://" + strApiHost + ":" + strApiPort;
        System.out.println("URL:" + strURL);
        return "redirect:" + strURL;
    }
}
```

3.6.3 运行程序

运行程序后,在浏览器中输入 localhost:8080/test/http/www.163.com/80,控制台的输出结果如图 3-18 所示;在浏览器中页面成功跳转到网易首页(https://www.163.com/)。

```
协议: http
主机: www.163.com
端口: 80
URL: http://www.163.com:80
```

图 3-18 在浏览器中输入 localhost:8080/test/http/www.163.com/80 后控制台的输出结果

3.7 用线程模拟区块链的示例

视频讲解

区块链应用往往会在不同机器(不同节点)上运行。本节通过不同线程来模拟不同机器上的区块链应用,由此可以体会到挖矿的时间先后关系。

3.7.1 创建项目并添加依赖

用 IDEA 创建完项目 beginex 之后,确保在文件 pom.xml 的 < dependencies > 和 </dependencies > 之间添加了 Web、Gson、Lombok 等依赖,代码如例 3-29 所示。

【例 3-29】 添加 Web、Gson、Lombok 依赖的代码示例。

```
<dependency>
    <groupId>org.springframework.boot</groupId>
    <artifactId>spring-boot-starter-web</artifactId>
</dependency>
<dependency>
    <groupId>com.google.code.gson</groupId>
    <artifactId>gson</artifactId>
    <version>2.8.5</version>
</dependency>
<dependency>
    <groupId>org.projectlombok</groupId>
    <artifactId>lombok</artifactId>
    <optional>true</optional>
</dependency>
```

3.7.2 创建类 Block

在包 com.bookcode 中创建 entity 子包,并在包 com.bookcode.entity 中创建类 Block,代码如例 3-30 所示。

【例 3-30】 类 Block 的代码示例。

```
package com.bookcode.entity;
import com.bookcode.utils.MineThread;
import com.bookcode.utils.OutWaitThread;
import com.bookcode.utils.StringUtil;
```

```
import lombok.Data;
import java.util.Date;
@Data
public class Block {
    private String hash;
    private String previousHash;
    private String data;
    private long timestamp;
    private int nonce;
    public Block(String data, String previousHash) {
        super();
        this.previousHash = previousHash;
        this.data = data;
        this.timestamp = new Date().getTime();
        this.hash = calculateHash();
    }
    public String calculateHash() {
        String calculatedHash = StringUtil.applySHA256(previousHash + timestamp + data);
        return calculatedHash;
    }
    public void mineBlock(int difficulty) {
        //模拟挖矿
        for (int i = 0; i < difficulty; i++) {
            MineThread mineThread = new MineThread();
            mineThread.start();
        }
        //挖矿完成后报告挖矿的结果
        OutWaitThread outWaitThread = new OutWaitThread();
        outWaitThread.start();
        System.out.println("挖出一个新区块,其哈希值为: " + hash);
    }
}
```

3.7.3 创建类 StringUtil

在包 com.bookcode 中创建 utils 子包,并在包 com.bookcode.utils 中创建类 StringUtil,代码如例 3-31 所示。

【例 3-31】 类 StringUtil 的代码示例。

```
package com.bookcode.utils;
import java.nio.charset.StandardCharsets;
import java.security.MessageDigest;
import java.security.NoSuchAlgorithmException;
public class StringUtil {
    public static String applySHA256(String input) {
        StringBuffer hexString = null;
        try {
```

```
            MessageDigest messageDigest = MessageDigest.getInstance("SHA-256");
            byte[] hash = messageDigest.digest(input.getBytes(StandardCharsets.UTF_8));
            hexString = new StringBuffer();
            String hex = null;
            for (int t = 0; t < hash.length; t++) {
                hex = Integer.toHexString(0xff & hash[t]);
                if (hex.length() == 1) {
                    hexString.append('0');
                } else {
                    hexString.append(hex);
                }
            }
        } catch (NoSuchAlgorithmException e) {
            e.printStackTrace();
        }
        return hexString.toString();
    }
}
```

3.7.4 创建类 MineThread

在包 com.bookcode.utils 中创建类 MineThread,代码如例 3-32 所示。

【例 3-32】 类 MineThread 的代码示例。

```
package com.bookcode.utils;
public class MineThread extends Thread {
    @Override
    public void run() {
        try {
System.out.println("挖矿的模拟线程为: " + currentThread().getName() + ",模拟挖矿的开始时间 = " + System.currentTimeMillis());
            Thread.sleep(3000);                    //延时 3s
System.out.println("挖矿的模拟线程为: " + currentThread().getName() + ",模拟挖矿的结束时间 = " + System.currentTimeMillis());
        } catch (InterruptedException e) {
            e.printStackTrace();
        }
    }
}
```

3.7.5 创建类 OutWaitThread

在包 com.bookcode.utils 中创建类 OutWaitThread,代码如例 3-33 所示。

【例 3-33】 类 OutWaitThread 的代码示例。

```
package com.bookcode.utils;
public class OutWaitThread extends Thread {
    @Override
    public void run() {
        try {
            Thread.sleep(1000); //延时1s
        } catch (InterruptedException e) {
            e.printStackTrace();
        }
    }
}
```

3.7.6 创建类 MineController

在包 com.bookcode 中创建 controller 子包，并在包 com.bookcode.controller 中创建类 MineController，代码如例 3-34 所示。

【例 3-34】 类 MineController 的代码示例。

```
package com.bookcode.controller;
import com.bookcode.entity.Block;
import com.google.gson.GsonBuilder;
import org.springframework.web.bind.annotation.GetMapping;
import org.springframework.web.bind.annotation.RestController;
import java.util.ArrayList;
@RestController
public class MineController {
    ArrayList<Block> blockchain = new ArrayList<>();
    int difficulty = 1;
    @GetMapping("/mine")
    public String testMine() {
        Long startTime = System.currentTimeMillis();
        blockchain.add(new Block("您好,我是第一个区块.", "0"));
        blockchain.get(0).mineBlock(difficulty);
blockchain.add(new Block("您好吗?我是第二个区块.", blockchain.get(blockchain.size()-1).
getHash()));
        blockchain.get(1).mineBlock(difficulty);
blockchain.add(new Block("好久不见,我是第三个区块.", blockchain.get(blockchain.size()-1).
getHash()));
        blockchain.get(2).mineBlock(difficulty);
        System.out.println(isChainValid()?"所挖的区块有效\n":"所挖的区块无效\n");
        String blockchainJson = new GsonBuilder().setPrettyPrinting().create().toJson
(blockchain);
        Long endTime = System.currentTimeMillis();
String stringResult = "当难度系数difficulty = " + difficulty + "时,挖矿共花掉" +
(endTime - startTime) + "毫秒.";
        return "所有区块信息:\n" + blockchainJson + "\n" + stringResult;
    }
```

```java
    private boolean isChainValid() {
        Block currentBlock;
        Block previousBlock;
        String hashTarget = new String(new char[difficulty]).replace('\0', '0');
        for (int i = 1; i < blockchain.size(); i++) {
            currentBlock = blockchain.get(i);
            previousBlock = blockchain.get(i - 1);
            if (!currentBlock.getHash().equals(currentBlock.calculateHash())) {
                System.out.println("计算出的区块 Hash 值与区块链中存储的值不相等.");
                return false;
            }
            if (!previousBlock.getHash().equals(currentBlock.getPreviousHash())) {
                System.out.println("计算出的区块的前一区块 Hash 值与区块链中存储的值不相等.");
                return false;
            }
            if (!currentBlock.getHash().substring(0, difficulty).equals(hashTarget)) {
                System.out.println("\n该区块已经被其他人(机器)挖出......");
                return false;
            }
        }
        return true;
    }
}
```

3.7.7 创建类 BlockController

在包 com.bookcode.controller 中创建类 BlockController，代码如例 3-35 所示。

【例 3-35】 类 BlockController 的代码示例。

```java
package com.bookcode.controller;
import com.bookcode.entity.Block;
import org.springframework.web.bind.annotation.GetMapping;
import org.springframework.web.bind.annotation.RestController;
import java.util.LinkedList;
import java.util.List;
@RestController
public class BlockController {
    @GetMapping("/block")
    public List<String> testBlock() {
        List<String> stringList = new LinkedList<String>();
        stringList.add("创建的所有区块信息:");
        Block genesisBlock = new Block("我是第一个(创世)区块", "0");
        System.out.println("第一个区块(创世区块)的 Hash 值:" + genesisBlock.calculateHash());
        System.out.println("第一个区块(创世区块)的信息:" + genesisBlock.toString());
        stringList.add(genesisBlock.toString());
        //请注意 secondBlock 与 genesisBlock 之间 Hash 值的关系
```

```
            Block secondBlock = new Block("我是第二个区块", genesisBlock.calculateHash());
            System.out.println("第二个区块的 Hash 值:" + secondBlock.calculateHash());
            System.out.println("第二个区块的信息:" + secondBlock.toString());
            stringList.add(secondBlock.toString());
            Block thirdBlock = new Block("我是第三个区块", secondBlock.calculateHash());
            System.out.println("第三个区块的 Hash 值:" + thirdBlock.calculateHash());
            System.out.println("第三个区块的信息:" + thirdBlock.toString());
            stringList.add(thirdBlock.toString());
            return stringList;
    }
}
```

3.7.8 创建类 BlockchainController

在包 com.bookcode.controller 中创建类 BlockchainController,代码如例 3-36 所示。

【例 3-36】 类 BlockchainController 的代码示例。

```
package com.bookcode.controller;
import com.bookcode.entity.Block;
import com.google.gson.GsonBuilder;
import org.springframework.web.bind.annotation.GetMapping;
import org.springframework.web.bind.annotation.RestController;
import java.util.ArrayList;
@RestController
public class BlockchainController {
    ArrayList<Block> blockChain = new ArrayList<>();
    @GetMapping("/blockchain")
    public String testBlockchain() {
        blockChain.add(new Block("第一个区块", "0"));
blockChain.add(new Block("第二个区块", blockChain.get(blockChain.size() - 1).calculateHash()));
blockChain.add(new Block("第三个区块", blockChain.get(blockChain.size() - 1).calculateHash()));
blockChain.add(new Block("第四个区块", blockChain.get(blockChain.size() - 1).calculateHash()));
String blockChainInJSON = new GsonBuilder().setPrettyPrinting().create().toJson(blockChain);
        System.out.println("普通格式的区块链信息:\n" + blockChain);
        System.out.println();
        System.out.println("检查区块链的有效性");
     System.out.println("是否是有效区块链:" + (isValidBlockChain(blockChain) ? "有效" : "无效"));
        System.out.println(" ========= ");
//注意对比前一个检查,体会区块链的不可篡改性(任何修改都会导致新区块和原来区块的不一致)
        System.out.println("检查修改后的区块的有效性");
```

```java
            blockChain.get(1).setHash("qwerty");
            blockChain.get(1).setPreviousHash("asdfg");
    System.out.println("是否是有效区块链：" + (isValidBlockChain(blockChain) ? "有效" : "无效"));
            return "JSON 格式的区块链信息：\n" + blockChainInJSON;
    }
    private boolean isValidBlockChain(ArrayList<Block> blockChain2) {
        Block currentBlock;
        Block previousBlock;
        for (int u = 1; u < blockChain2.size(); u++) {
            currentBlock = blockChain2.get(u);
            previousBlock = blockChain2.get(u - 1);
            if (!currentBlock.getHash().equals(currentBlock.calculateHash())) {
                System.out.println("本区块的当前 Hash 值不相等,是无效区块.");
                return false;
            }
            if (!previousBlock.getHash().equals(currentBlock.getPreviousHash())) {
                System.out.println("本区块的前一个区块 Hash 值不相等,是无效区块.");
                return false;
            }
        }
        return true;
    }
}
```

3.7.9 运行程序

运行程序后,在浏览器中输入 localhost:8080/mine 后浏览器的输出结果如图 3-19 所示,控制台的输出结果如图 3-20 所示。请注意图 3-20 中三个模拟线程的挖矿开始时间、结束时间存在差异,并由此确定挖矿是否有效;并请注意,线程编号是一直在增加的(即使是重新运行程序也不会从 0 开始),这模拟了区块链信息的不可篡改性。在浏览器中输入 localhost:8080/block 后,浏览器的输出结果如图 3-21 所示,控制台的输出结果如图 3-22 所示。在浏览器中输入 localhost:8080/blockchain 后,浏览器的输出结果如图 3-23 所示,控制台的输出结果如图 3-24 所示。

图 3-19　在浏览器中输入 localhost:8080/mine 后浏览器的输出结果

第3章 区块链应用的基础开发

```
挖矿的模拟线程为：Thread-16，模拟挖矿的开始时间=1561330256695
挖出一个新区块，其哈希值为：d28c43c083f4e6fa00e91b70a32154ebb8ae88a5c2168c4d0b7b2e6793a27
挖出一个新区块，其哈希值为：6a48d3933abbbccea8c5f2803952d0c5c7041bfc2cc4f97771039f958a07717
挖出一个新区块，其哈希值为：f8d0ab538b529fa89ae0caa91e18b3b8bdc96c43d955cdf4160f20203c4484

该区块已经被其他人（机器）挖出……
所挖的区块无效

挖矿的模拟线程为：Thread-18，模拟挖矿的开始时间=1561330256697
挖矿的模拟线程为：Thread-20，模拟挖矿的开始时间=1561330256698
挖矿的模拟线程为：Thread-16，模拟挖矿的结束时间=1561330259696
挖矿的模拟线程为：Thread-20，模拟挖矿的结束时间=1561330259699
挖矿的模拟线程为：Thread-18，模拟挖矿的结束时间=1561330259699
```

图 3-20　在浏览器中输入 localhost:8080/mine 后控制台的输出结果

```
["创建的所有区块信息：","Block(hash=fdfc0377d415939b4b1fdb67293d8559d48b705
previousHash=0, data=我是第一个（创世）区块, timestamp=1578989948492,
nonce=0)","Block(hash=ae7e06cc0e1e103657d53fb86e6e769aa686148afeab12bac4120
previousHash=fdfc0377d415939b4b1fdb67293d8559d48b70576f7f5b5e8f05c2021cffb,
timestamp=1578989948493,
nonce=0)","Block(hash=4190c45098b8ece863e4d56fef661ab0d277aa115ca3e132703e8
previousHash=ae7e06cc0e1e103657d53fb86e6e769aa686148afeab12bac412006cf8e0,
timestamp=1578989948493, nonce=0)"]
```

图 3-21　在浏览器中输入 localhost:8080/block 后浏览器的输出结果

```
第一个区块（创世区块）的Hash值：fdfc0377d415939b4b1fdb67293d8559d48b70576f7f5b5e8f05c2021cffb
第一个区块（创世区块）的信息：Block(hash=fdfc0377d415939b4b1fdb67293d8559d48b70576f7f5b5e8f05c2021cffb, previousHas
第二个区块的Hash值：ae7e06cc0e1e103657d53fb86e6e769aa686148afeab12bac412006cf8e0
第二个区块的信息：Block(hash=ae7e06cc0e1e103657d53fb86e6e769aa686148afeab12bac412006cf8e0, previousHash=fdfc0377d4
第三个区块的Hash值：4190c45098b8ece863e4d56fef661ab0d277aa115ca3e132703e86197eac071
第三个区块的信息：Block(hash=4190c45098b8ece863e4d56fef661ab0d277aa115ca3e132703e86197eac071, previousHash=ae7e06c
```

图 3-22　在浏览器中输入 localhost:8080/block 后控制台的输出结果

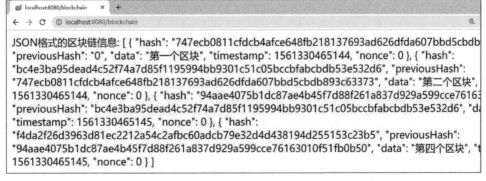

图 3-23　在浏览器中输入 localhost:8080/blockchain 后浏览器的输出结果

```
普通格式的区块链信息:
[Block(hash=747ecb0811cfdcb4afce648fb2

检查区块链的有效性
是否是有效区块链: 有效
=========
检查修改后的区块的有效性
本区块的当前Hash值不相等,是无效区块。
是否是有效区块链: 无效
```

图 3-24 在浏览器中输入 localhost:8080/blockchain 后控制台的输出结果

习题 3

实验题

1. 模拟实现两个钱包之间的交互。
2. 模拟实现挖矿功能。
3. 模拟实现创建钱包和利用钱包进行交易。
4. 简要实现 Merkle 树。
5. 模拟实现 Gossip 协议。
6. 实现 IOTA 的模拟应用。

第4章 区块链应用的P2P实现

狭义地讲，P2P 是指 P2P 网络；广义地讲，P2P 是指基于 P2P 形式的不同实现，可以在网络层，也可以在其他非网络层。本章的 P2P 实现主要是指广义的 P2P 实现。本章介绍如何基于 Java-WebSocket 实现 P2P 网络、基于 WebSocket 实现 P2P 页面互连、基于 t-io 实现 P2P 网络、基于 ZooKeeper 实现 P2P 服务、基于 Web Service 和 CXF 实现 P2P 服务。

4.1 基于 Java-WebSocket 实现 P2P 网络

视频讲解

本节介绍如何基于 Java-WebSocket 实现 P2P 网络，模拟服务器端和客户端连接后互发消息。

4.1.1 创建项目并添加依赖

用 IDEA 创建完项目 websocketp2p 之后，确保在文件 pom.xml 的<dependencies>和</dependencies>之间添加了 Java-WebSocket 和 Lombok 依赖，代码如例 4-1 所示。

【例 4-1】 添加 Java-WebSocket 和 Lombok 依赖的代码示例。

```
<dependency>
    <groupId>org.java-websocket</groupId>
    <artifactId>Java-WebSocket</artifactId>
    <version>1.3.8</version>
</dependency>
<dependency>
    <groupId>org.projectlombok</groupId>
    <artifactId>lombok</artifactId>
    <version>1.18.4</version>
</dependency>
```

4.1.2 创建类 P2PUtil

在包 com.bookcode 中创建 util 子包,并在包 com.bookcode.util 中创建类 P2PUtil,代码如例 4-2 所示。

【例 4-2】 类 P2PUtil 的代码示例。

```java
package com.bookcode.util;
import lombok.Data;
import org.apache.logging.log4j.util.Strings;
import org.java_websocket.WebSocket;
import java.util.ArrayList;
import java.util.List;
@Data
public class P2PUtil {
    public static List<WebSocket> ls = new ArrayList<WebSocket>();
    public static void sendMessage(WebSocket webSocket, String s) {
        System.out.println("给" + webSocket.getRemoteSocketAddress().getPort() + "的 P2P 消息是:" + s);
        webSocket.send(s);
    }
    public static void broadcast(String message) throws Exception {
        if (ls.size() == 0 || Strings.isNotEmpty(message)){
            return;
        }
        System.out.println("开始广播!");
        for (WebSocket item : ls) {
            try {
                sendMessage((WebSocket) item,message);
            } catch (Exception e) {
                continue;
            }
        }
        System.out.println("广播结束!");
    }
}
```

4.1.3 创建类 P2Pserver

在包 com.bookcode 中创建 service 子包,并在包 com.bookcode.service 中创建类 P2Pserver,代码如例 4-3 所示。

【例 4-3】 类 P2Pserver 的代码示例。

```java
package com.bookcode.service;
import com.bookcode.util.P2PUtil;
import org.java_websocket.handshake.ClientHandshake;
```

```java
import org.java_websocket.server.WebSocketServer;
import org.springframework.core.annotation.Order;
import org.springframework.stereotype.Component;
import org.java_websocket.WebSocket;
import javax.annotation.PostConstruct;
import java.net.InetSocketAddress;
@Component
public class P2Pserver {
    private int port = 8001;
    @PostConstruct
    @Order(1)
    public void init(){
        final WebSocketServer ss = new WebSocketServer(new InetSocketAddress(port)) {
            @Override
            public void onOpen(WebSocket webSocket, ClientHandshake clientHandshake) {
                P2PUtil.sendMessage(webSocket,"服务器端成功创建连接!");
                P2PUtil.ls.add(webSocket);
            }
            @Override
            public void onClose(WebSocket webSocket, int i, String s, boolean b) {
                System.out.println("与客户端" + webSocket.getRemoteSocketAddress() + "的连接关闭!");
                P2PUtil.ls.remove(webSocket);
            }
            @Override
            public void onMessage(WebSocket webSocket, String s) {
                P2PUtil.sendMessage(webSocket,"服务器端收到消息!");
            }
            @Override
            public void onError(WebSocket webSocket, Exception e) {
                System.out.println("与客户端" + webSocket.getRemoteSocketAddress() + "的连接出错!");
                P2PUtil.ls.remove(webSocket);
            }
            @Override
            public void onStart() {
                System.out.println("服务器端启动!");
            }
        };
        ss.start();
        System.out.println("服务器端开始监听端口:" + port + "!");
    }
}
```

4.1.4 创建类 P2Pclient

在包 com.bookcode.service 中创建类 P2Pclient,代码如例 4-4 所示。

【例 4-4】 类 P2Pclient 的代码示例。

```java
package com.bookcode.service;
import com.bookcode.util.P2PUtil;
import org.java_websocket.client.WebSocketClient;
import org.java_websocket.handshake.ServerHandshake;
import org.springframework.core.annotation.Order;
import org.springframework.stereotype.Component;
import javax.annotation.PostConstruct;
import java.net.URI;
import java.net.URISyntaxException;
@Component
public class P2Pclient {
    private String url = "ws://localhost:8001/";
    @PostConstruct
    @Order(2)
    public void conServer(){
        try{
            final WebSocketClient sc = new WebSocketClient(new URI(url)) {
                @Override
                public void onOpen(ServerHandshake serverHandshake) {
                    P2PUtil.sendMessage(this,"客户端创建成功!");
                    P2PUtil.ls.add(this);
                }
                @Override
                public void onMessage(String s) {
                    System.out.println("客户端向服务器端发送的消息是:" + s);
                }
                @Override
                public void onClose(int i, String s, boolean b) {
                    System.out.println("客户端关闭!");
                    P2PUtil.ls.remove(this);
                }
                @Override
                public void onError(Exception e) {
                    System.out.println("客户端出错!");
                    P2PUtil.ls.remove(this);
                }
            };
            sc.connect();
        } catch (URISyntaxException e){
            System.out.println("连接出错:" + e.getMessage());
        }
    }
}
```

4.1.5 运行程序

运行程序后,控制台的输出结果如图 4-1 所示。图 4-1 中内容空白的矩形框是用于覆盖程序运行的过程信息,增加矩形框的目的是更好地展示控制台中文字输出结果而做的处理。

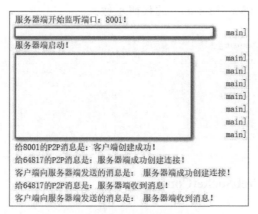

图 4-1 控制台中的输出结果

4.2 基于 WebSocket 实现 P2P 页面互连

视频讲解

本节介绍如何基于 WebSocket 实现 P2P 页面互连,模拟钱包的发送和历史信息。

4.2.1 添加依赖

在 4.1 节项目 websocketp2p 的基础上,向文件 pom.xml 中< dependencies >和</dependencies>之间添加 WebSocket、thymeleaf 和 Web 依赖,代码如例 4-5 所示。请注意这里的 WebSocket 依赖和 4.1.1 节 Java-WebSocket 依赖的不同。

【例 4-5】 添加 WebSocket、thymeleaf 和 Web 依赖的代码示例。

```
< dependency >
    < groupId > org.springframework.boot </groupId >
    < artifactId > spring - boot - starter - websocket </artifactId >
</dependency >
< dependency >
    < groupId > org.springframework.boot </groupId >
    < artifactId > spring - boot - starter - thymeleaf </artifactId >
</dependency >
< dependency >
    < groupId > org.springframework.boot </groupId >
    < artifactId > spring - boot - starter - web </artifactId >
</dependency >
```

4.2.2 创建类 WalletSendMessage

在包 com.bookcode 中创建 entity 子包,并在包 com.bookcode.entity 中创建类 WalletSendMessage,代码如例 4-6 所示。

【例 4-6】 类 WalletSendMessage 的代码示例。

```java
package com.bookcode.entity;
public class WalletSendMessage {
    public String message;
    public String tBC;
    public String date;
}
```

4.2.3　创建类 WebSocketConfig

在包 com.bookcode 中创建 config 子包,并在包 com.bookcode.config 中创建类 WebSocketConfig,代码如例 4-7 所示。

【例 4-7】 类 WebSocketConfig 的代码示例。

```java
package com.bookcode.config;
import org.springframework.context.annotation.Configuration;
import org.springframework.messaging.simp.config.MessageBrokerRegistry;
import org.springframework.web.socket.config.annotation.EnableWebSocketMessageBroker;
import org.springframework.web.socket.config.annotation.StompEndpointRegistry;
import org.springframework.web.socket.config.annotation.WebSocketMessageBrokerConfigurer;
@Configuration
@EnableWebSocketMessageBroker
public class WebSocketConfig implements WebSocketMessageBrokerConfigurer {
    @Override
    public void configureMessageBroker(MessageBrokerRegistry config) {
        config.enableSimpleBroker("/topic");
        config.setApplicationDestinationPrefixes("/app");
    }
    @Override
    public void registerStompEndpoints(StompEndpointRegistry registry) {
        registry.addEndpoint("/my-websocket").withSockJS();
    }
}
```

4.2.4　创建类 WalletController

在包 com.bookcode 中创建 controller 子包,并在包 com.bookcode.controller 中创建类 WalletController,代码如例 4-8 所示。

【例 4-8】 类 WalletController 的代码示例。

```java
package com.bookcode.controller;
import com.bookcode.entity.WalletSendMessage;
import org.springframework.beans.factory.annotation.Autowired;
```

```java
import org.springframework.messaging.handler.annotation.MessageMapping;
import org.springframework.messaging.handler.annotation.SendTo;
import org.springframework.messaging.simp.SimpMessagingTemplate;
import org.springframework.scheduling.annotation.EnableScheduling;
import org.springframework.scheduling.annotation.Scheduled;
import org.springframework.stereotype.Controller;
import org.springframework.web.bind.annotation.GetMapping;
import java.text.DateFormat;
import java.text.SimpleDateFormat;
import java.util.Date;
@EnableScheduling
@Controller
public class WalletController {
    @Autowired
    private SimpMessagingTemplate messagingTemplate;
    @GetMapping("/")
    public String index() {
        return "index";
    }
    @MessageMapping("/send")
    @SendTo("/topic/send")
    public WalletSendMessage send(WalletSendMessage message) {
        DateFormat df = new SimpleDateFormat("yyyy-MM-dd HH:mm:ss");
        message.date = df.format(new Date());
        return message;
    }
    @Scheduled(fixedRate = 1000)
    @SendTo("/topic/callback")
    public Object callback() {
        DateFormat df = new SimpleDateFormat("yyyy-MM-dd HH:mm:ss");
        messagingTemplate.convertAndSend("/topic/callback", df.format(new Date()));
        return "callback";
    }
}
```

4.2.5 创建文件 index.html

在目录 src/main/resources/templates 下,创建文件 index.html 并修改文件代码,修改后的文件代码如例 4-9 所示。一般说来,创建 HTML 文件并修改文件代码的过程简称为创建 HTML 文件,对应的文件代码也默认为修改后的文件代码。后面章节也遵守此约定。

【例 4-9】 修改后的文件 index.html 代码示例。

```html
<!DOCTYPE html>
<html>
<head>
    <title>Spring Boot—WebSocket 页面互连</title>
    <script src="https://cdn.bootcss.com/angular.js/1.7.7/angular.min.js"></script>
    <script src="https://cdn.bootcss.com/sockjs-client/1.3.0/sockjs.min.js"></script>
    <script src="https://cdn.bootcss.com/stomp.js/2.3.3/stomp.min.js"></script>
```

```html
<script type="text/javascript">
    var stompClient = null;
    var app = angular.module('app', []);
    app.controller('MainController', function($rootScope, $scope, $http) {
        $scope.data = {
            //连接状态
            connected : false,
            //消息
            message : '',
            rows : []
        };
        //连接
        $scope.connect = function() {
            var socket = new SockJS('/my-websocket');
            stompClient = Stomp.over(socket);
            stompClient.connect({}, function(frame) {
                //注册发送消息
                stompClient.subscribe('/topic/send', function(msg) {
                    $scope.data.rows.push(JSON.parse(msg.body));
                    $scope.data.connected = true;
                    $scope.$apply();
                });
                //注册推送时间回调
                stompClient.subscribe('/topic/callback', function(r) {
                    $scope.data.time = '当前服务器时间: ' + r.body;
                    $scope.data.connected = true;
                    $scope.$apply();
                });
                $scope.data.connected = true;
                $scope.$apply();
            });
        };
        $scope.disconnect = function() {
            if (stompClient != null) {
                stompClient.disconnect();
            }
            $scope.data.connected = false;
        }
        $scope.send = function() {
            stompClient.send("/app/send", {}, JSON.stringify({
                'message' : $scope.data.message,
                'tBC' : $scope.data.tBC
            }));
        }
    });
</script>
</head>
<body ng-app="app" ng-controller="MainController">
<h2>Spring Boot—WebSocket 页面互连</h2>
```

```html
<button type = "button" ng - disabled = "data.connected" ng - click = "connect()">连接
</button>
<button type = "button" ng - click = "disconnect()"
        ng - disabled = "!data.connected">断开</button>
<br />
<br />
<div ng - show = "data.connected">
    <label>{{data.time}}</label><br /><br />
<input type = "text" ng - model = "data.message" placeholder = "请输入接收地址..." />
    <br />
<input type = "text" ng - model = "data.tBC" placeholder = "请输入发送比特币数..." />
    <button ng - click = "send()" type = "button">发送</button>
    <br /><br />历史:<br />
    <table>
        <thead>
        <tr>
            <th>地址信息</th>
            <th>发送比特币数</th>
            <th>发送时间</th>
        </tr>
        </thead>
        <tbody>
        <tr ng - repeat = "row in data.rows">
            <td>{{row.message}}</td>
            <td>{{row.tBC}}</td>
            <td>{{row.date}}</td>
        </tr>
        </tbody>
    </table>
</div>
</body>
</html>
```

4.2.6 运行程序

运行程序后,打开第一个浏览器窗口;在浏览器中输入 localhost:8080 后单击"连接"按钮,再在浏览器中输入模拟钱包 0002 发送 2 比特币的信息后单击"发送"按钮,浏览器中"历史"下面显示一条钱包发送比特币后的模拟信息"第 1 行信息",第一个浏览器窗口的结果如图 4-2 所示。再打开第二个浏览器窗口,输入模拟钱包 0001 发送 1 比特币的信息(第 2 行信息)后单击"发送"按钮,第二个浏览器窗口的结果如图 4-3 所示,与此同时,第一个浏览器窗口的结果如图 4-4 所示。对比图 4-2、图 4-3 和图 4-4 可以发现,互连的 P2P 页面中历史信息同步更新,只要是任何一个页面中发送比特币,两个页面中都会记录下(并显示)最新的发送比特币信息。

图 4-2　在第一个浏览器窗口中输入 localhost:8080 并发送一条信息后第一个浏览器窗口的结果

图 4-3　在第二个浏览器窗口中输入 localhost:8080 并发送一条信息后第二个浏览器窗口的结果

图 4-4　在第二个浏览器窗口中输入 localhost:8080 并发送一条信息后第一个浏览器窗口的结果

4.3　基于 t-io 实现 P2P 网络

视频讲解

　　t-io 是一个网络框架,为常见和网络相关的业务(如消息推送、RPC、监控)提供了较完整的解决方案,即丰富的编程 API,极大地减少业务层的编程难度。

4.3.1 添加依赖

在 4.2 节项目 websocketp2p 的基础上，向文件 pom.xml 中 < dependencies > 和 </dependencies > 之间添加了 tio-core 依赖，代码如例 4-10 所示。

【例 4-10】 添加 tio-core 依赖的代码示例。

```
<dependency>
        <groupId>org.t-io</groupId>
        <artifactId>tio-core</artifactId>
        <version>3.2.6.v20190110-RELEASE</version>
</dependency>
```

4.3.2 创建类 ServerPacket

在包 com.bookcode.entity 中创建类 ServerPacket，代码如例 4-11 所示。

【例 4-11】 类 ServerPacket 的代码示例。

```
package com.bookcode.entity;
import lombok.Data;
import org.tio.core.intf.Packet;
@Data
public class ServerPacket extends Packet {
    public static final Integer PACKET_HEADER_LENGTH = 4;
    public static final Integer PORT = 8999;
    byte[] body; //数据
}
```

4.3.3 创建类 MServerAioHandler

在包 com.bookcode.util 中创建类 MServerAioHandler，代码如例 4-12 所示。

【例 4-12】 类 MServerAioHandler 的代码示例。

```
package com.bookcode.util;
import com.bookcode.entity.ServerPacket;
import org.slf4j.Logger;
import org.slf4j.LoggerFactory;
import org.tio.core.ChannelContext;
import org.tio.core.GroupContext;
import org.tio.core.Node;
import org.tio.core.Tio;
import org.tio.core.exception.AioDecodeException;
import org.tio.core.intf.Packet;
```

```java
import org.tio.server.intf.ServerAioHandler;
import java.nio.ByteBuffer;
public class MServerAioHandler implements ServerAioHandler {
    private static final Logger logger = LoggerFactory.getLogger(MServerAioHandler.class);
    public Packet decode(ByteBuffer byteBuffer, int limit, int position, int readableLength, ChannelContext channelContext) throws AioDecodeException {
        logger.debug("inside decode...");
        if (readableLength < ServerPacket.PACKET_HEADER_LENGTH) {
            return null;
        }
        int bodyLength = byteBuffer.getInt();
        if (bodyLength < 0) {
             throw new AioDecodeException("body length[" + bodyLength + "] is invalid. romote: " + channelContext.getServerNode());
        }
        int len = bodyLength + ServerPacket.PACKET_HEADER_LENGTH;
        if (len > readableLength) {
            return null;
        } else {
            byte[] bytes = new byte[len];
            int i = 0;
            while(true){
                if(byteBuffer.remaining() == 0){
                    break;
                }
                byte b = byteBuffer.get();
                bytes[i++] = b;
            }
            ServerPacket serverPacket = new ServerPacket();
            serverPacket.setBody(bytes);
            return serverPacket;
        }
    }
    public ByteBuffer encode (Packet packet, GroupContext groupContext, ChannelContext channelContext) {
        logger.info("inside encode...");
        ServerPacket serverPacket = (ServerPacket) packet;
        byte[] body = serverPacket.getBody();
        int bodyLength = 0;
        if(body != null){
            bodyLength = body.length;
        }
         ByteBuffer byteBuffer = ByteBuffer.allocate(bodyLength + ServerPacket.PACKET_HEADER_LENGTH);
        byteBuffer.order(groupContext.getByteOrder());
        byteBuffer.putInt(bodyLength);
        if (body != null) {
            byteBuffer.put(body);
        }
        return byteBuffer;
```

```
    }
    public void handler(Packet packet, ChannelContext channelContext) throws Exception {
        logger.debug("inside handler...");
        channelContext.setServerNode(new Node("127.0.0.1", ServerPacket.PORT));
        ServerPacket serverPacket = (ServerPacket) packet;
        byte[] body = serverPacket.getBody();
        if(body != null){
            String bodyStr = new String(body, "utf-8");
            ServerPacket serverPacket1 = new ServerPacket();
serverPacket1.setBody(("receive from [" + channelContext.getClientNode() + "]: " + bodyStr).getBytes("utf-8"));
            Tio.send(channelContext, serverPacket1);
        }
    }
}
```

4.3.4 创建类 MServerAioListener

在包 com.bookcode.util 中创建类 MServerAioListener，代码如例 4-13 所示。

【例 4-13】 类 MServerAioListener 的代码示例。

```
package com.bookcode.util;
import org.tio.core.ChannelContext;
import org.tio.core.intf.Packet;
import org.tio.server.intf.ServerAioListener;
public class MServerAioListener implements ServerAioListener {
    public void onAfterConnected(ChannelContext channelContext, boolean b, boolean b1) throws Exception {
    }
    public void onAfterDecoded(ChannelContext channelContext, Packet packet, int i) throws Exception {
    }
    public void onAfterReceivedBytes(ChannelContext channelContext, int i) throws Exception {
    }
    public void onAfterSent(ChannelContext channelContext, Packet packet, boolean b) throws Exception {
    }
    public void onAfterHandled(ChannelContext channelContext, Packet packet, long l) throws Exception {
    }
    public void onBeforeClose(ChannelContext channelContext, Throwable throwable, String s, boolean b) throws Exception {
    }
}
```

4.3.5 创建类 MClientAioHandler

在包 com.bookcode.util 中创建类 MClientAioHandler，代码如例 4-14 所示。

【例 4-14】 类 MClientAioHandler 的代码示例。

```
package com.bookcode.util;
import com.bookcode.entity.ServerPacket;
import org.slf4j.Logger;
import org.slf4j.LoggerFactory;
import org.tio.client.intf.ClientAioHandler;
import org.tio.core.ChannelContext;
import org.tio.core.GroupContext;
import org.tio.core.exception.AioDecodeException;
import org.tio.core.intf.Packet;
import java.nio.ByteBuffer;
public class MClientAioHandler implements ClientAioHandler {
    Logger logger = LoggerFactory.getLogger(MClientAioHandler.class);
    @Override
    public Packet decode(ByteBuffer byteBuffer, int limit, int position, int readableLength, ChannelContext channelContext) throws AioDecodeException {
        if(readableLength < ServerPacket.PACKET_HEADER_LENGTH){
            return null;
        }
        int bodyLength = byteBuffer.getInt();
        if(bodyLength < 0){
            throw new AioDecodeException("body length is invalid. romote: " + channelContext.getServerNode());
        }
        int usefulLength = ServerPacket.PACKET_HEADER_LENGTH + bodyLength;
        if(usefulLength > readableLength){
            return null;
        }else {
            ServerPacket packet = new ServerPacket();
            byte[] body = new byte[bodyLength];
            byteBuffer.get(body);
            packet.setBody(body);
            return packet;
        }
    }
    @Override
    public ByteBuffer encode(Packet packet, GroupContext groupContext, ChannelContext channelContext) {
        ServerPacket clientPacket = (ServerPacket) packet;
        byte[] body = clientPacket.getBody();
        int bodyLength = 0;
        if(body != null){
            bodyLength = body.length;
```

```
            }
            int len = ServerPacket.PACKET_HEADER_LENGTH + bodyLength;
            ByteBuffer byteBuffer = ByteBuffer.allocate(len);
            byteBuffer.order(groupContext.getByteOrder());
            byteBuffer.putInt(bodyLength);
            if(body != null){
                byteBuffer.put(body);
            }
            return byteBuffer;
        }
        @Override
        public void handler(Packet packet, ChannelContext channelContext) throws Exception {
            ServerPacket clientPacket = (ServerPacket) packet;
            byte[] body = clientPacket.getBody();
            if(body != null){
                String bodyStr = new String(body, "utf-8");
                logger.debug("客户端收到消息: " + bodyStr);
            }
        }
        @Override
        public Packet heartbeatPacket(ChannelContext channelContext) {
            return null;
        }
    }
```

4.3.6 创建类 MClientAioListener

在包 com.bookcode.util 中创建类 MClientAioListener，代码如例 4-15 所示。

【例 4-15】 类 MClientAioListener 的代码示例。

```
package com.bookcode.util;
import com.bookcode.entity.ServerPacket;
import org.slf4j.Logger;
import org.slf4j.LoggerFactory;
import org.tio.client.intf.ClientAioListener;
import org.tio.core.ChannelContext;
import org.tio.core.Tio;
import org.tio.core.intf.Packet;
public class MClientAioListener implements ClientAioListener {
    Logger logger = LoggerFactory.getLogger(MClientAioListener.class);
    private static Integer count = 0;
    @Override
    public void onAfterConnected(ChannelContext channelContext, boolean b, boolean b1) throws Exception {
        logger.info("onAfterConnected!");
    }
    @Override
```

```java
    public void onAfterDecoded(ChannelContext channelContext, Packet packet, int i) throws Exception {
        logger.info("onAfterDecoded...");
    }
    @Override
    public void onAfterReceivedBytes(ChannelContext channelContext, int i) throws Exception {
        logger.info("onAfterReceivedBytes--------------------" + i);
    }
    @Override
    public void onAfterSent(ChannelContext channelContext, Packet packet, boolean b) throws Exception {
        logger.info("onAfterSent...");
    }
    @Override
    public void onAfterHandled(ChannelContext channelContext, Packet packet, long l) throws Exception {
        System.out.println("onAfterHandled...");
        ServerPacket clientPacket = (ServerPacket) packet;
        String resData = new String(clientPacket.getBody(), "utf-8");
        logger.info("[" + channelContext.getServerNode() + "]: " + resData);
        count++;
        ((ServerPacket) packet).setBody(("[" + channelContext.getServerNode() + "]: " + count).getBytes());
        Thread.sleep(5000);
        Tio.send(channelContext, packet);
    }
    @Override
    public void onBeforeClose(ChannelContext channelContext, Throwable throwable, String s, boolean b) throws Exception {
        logger.error(throwable.getMessage());
        logger.info(s);
    }
}
```

4.3.7 创建类 TIOServer

在包 com.bookcode.service 中创建类 TIOServer，代码如例 4-16 所示。

【例 4-16】 类 TIOServer 的代码示例。

```java
package com.bookcode.service;
import com.bookcode.util.MServerAioHandler;
import com.bookcode.util.MServerAioListener;
import org.springframework.stereotype.Component;
import org.tio.server.ServerGroupContext;
import org.tio.server.TioServer;
@Component
public class TIOServer {
```

```java
    public String startupTIO() {
        try{
            ServerGroupContext serverGroupContext = new ServerGroupContext("tio-server", new MServerAioHandler(), new MServerAioListener());
            TioServer server = new TioServer(serverGroupContext);
            TioServer tioServer = new TioServer(serverGroupContext);
                server.start("127.0.0.1", 8999);
        }catch (Exception e){
            System.out.println("出现异常: " + e.getMessage());
            return "error!";
        }
        return "Startup Server OK";
    }
}
```

4.3.8　创建类 TIOClient

在包 com.bookcode.service 中创建类 TIOClient，代码如例 4-17 所示。

【例 4-17】　类 TIOClient 的代码示例。

```java
package com.bookcode.service;
import com.bookcode.entity.ServerPacket;
import com.bookcode.util.MClientAioHander;
import com.bookcode.util.MClientAioListener;
import org.springframework.stereotype.Component;
import org.tio.client.ClientChannelContext;
import org.tio.client.ClientGroupContext;
import org.tio.client.TioClient;
import org.tio.core.Node;
@Component
public class TIOClient {
        public String startupTIO() {
            try {
ClientGroupContext clientGroupContext = new ClientGroupContext(new MClientAioHander(), new MClientAioListener());
                TioClient tioClient = new TioClient(clientGroupContext);
                System.out.println("tio 连接开始: ");
                ServerPacket clientPacket = new ServerPacket();
                clientPacket.setBody("hello,t-tio".getBytes("utf-8"));
ClientChannelContext clientChannelContext = tioClient.connect(new Node("127.0.0.1", 8999));
                clientPacket.setBody("hello,t-tio".getBytes("utf-8"));
                System.out.println("tio 连接关闭: ");
            }catch (Exception e){
                System.out.println("出现异常: " + e.getMessage());
                return "error!";
            }
```

```
        return "Startup Client OK";
    }
}
```

4.3.9 创建类 TIOController

在包 com.bookcode.controller 中创建类 TIOController,代码如例 4-18 所示。

【例 4-18】 类 TIOController 的代码示例。

```
package com.bookcode.controller;
import com.bookcode.service.TIOClient;
import com.bookcode.service.TIOServer;
import org.springframework.beans.factory.annotation.Autowired;
import org.springframework.web.bind.annotation.GetMapping;
import org.springframework.web.bind.annotation.RestController;
@RestController
public class TIOController {
    @Autowired
    private TIOServer tioServer;
    @Autowired
    private TIOClient tioClient;
    @GetMapping("/tioserver")
    public String server() {
        return tioServer.startupTIO();
    }
    @GetMapping("/tioclient")
    public String client() {
        return tioClient.startupTIO();
    }
}
```

4.3.10 运行程序

运行程序后,在浏览器中输入 localhost:8080/tioserver,控制台的输出结果如图 4-5 所示,浏览器的输出结果如图 4-6 所示。在浏览器中输入 localhost:8080/tioclient,将会创建一条从客户端到服务器的连接,浏览器的输出结果如图 4-7 所示。每次重新在浏览器中输入 localhost:8080/tioclient 都会创建一条连接(只要服务器没有关闭)。多次创建连接后控制台的输出结果如图 4-8 所示。运行过程中显示所有有效的连接如图 4-9 所示,关闭了一条连接(2 号连接)后显示所有有效的连接如图 4-10 所示。假如服务器正常运行时,再次在浏览器中输入 localhost:8080/tioserver,就会出错(请注意和多次输入 localhost:8080/tioclient 的不同);此时控制台的输出结果如图 4-11 所示,浏览器的输出结果如图 4-12 所示。

```
| Tio gitee address | https://gitee.com/tywo45/t-io       |
| Tio site address  | https://t-io.org/                   |
| Tio version       | 3.2.6.v20190110-RELEASE             |

| GroupContext name | tio-server                          |
| Started at        | 2019-02-10 14:36:32                 |
| Listen on         | 127.0.0.1:8999                      |
| Main Class        | java.lang.Thread                    |
| Jvm start time    | 14016 ms                            |
| Tio start time    | 8 ms                                |
| Pid               | 13316                               |
```

图 4-5　在浏览器中输入 localhost:8080/tioserver 后控制台的输出结果

图 4-6　在浏览器中输入 localhost:8080/tioserver 后浏览器的输出结果

图 4-7　在浏览器中输入 localhost:8080/tioclient 后浏览器的输出结果

```
tio连接开始:
2019-02-10 14:38:29.519  INFO 13316 --- [imer-heartbeat2] org.tio.client.TioClient              : [2]: curr:0, clos
2019-02-10 14:38:29.535  INFO 13316 --- [    tio-group-3] o.t.client.ConnectionCompletionHandler : connected to 127.
tio连接关闭:
2019-02-10 14:38:29.535  INFO 13316 --- [    tio-group-3] com.bookcode.util.MClientAioListener   : onAfterConnected!
tio连接开始:
2019-02-10 14:38:53.992  INFO 13316 --- [imer-heartbeat3] org.tio.client.TioClient              : [3]: curr:0, clos
2019-02-10 14:38:53.994  INFO 13316 --- [    tio-group-4] o.t.client.ConnectionCompletionHandler : connected to 127.
tio连接关闭:
2019-02-10 14:38:53.994  INFO 13316 --- [    tio-group-4] com.bookcode.util.MClientAioListener   : onAfterConnected!
tio连接开始:
2019-02-10 14:38:56.521  INFO 13316 --- [imer-heartbeat4] org.tio.client.TioClient              : [4]: curr:0, clos
2019-02-10 14:38:56.523  INFO 13316 --- [    tio-group-7] o.t.client.ConnectionCompletionHandler : connected to 127.
```

图 4-8　多次在浏览器中输入 localhost:8080/tioclient 后控制台的输出结果（建立多条连接）

```
[imer-heartbeat2] org.tio.client.TioClient        : [2]: curr:1, closed:0,
[imer-heartbeat3] org.tio.client.TioClient        : [3]: curr:1, closed:0,
[imer-heartbeat4] org.tio.client.TioClient        : [4]: curr:1, closed:0,
[imer-heartbeat5] org.tio.client.TioClient        : [5]: curr:1, closed:0,
[imer-heartbeat2] org.tio.client.TioClient        : [2]: curr:1, closed:0,
[imer-heartbeat3] org.tio.client.TioClient        : [3]: curr:1, closed:0,
[imer-heartbeat4] org.tio.client.TioClient        : [4]: curr:1, closed:0,
[imer-heartbeat5] org.tio.client.TioClient        : [5]: curr:1, closed:0,
[imer-heartbeat2] org.tio.client.TioClient        : [2]: curr:1, closed:0,
```

图 4-9　运行过程中控制台显示所有有效连接

```
org.tio.client.TioClient             : [2]: curr:0, closed:1,
org.tio.client.TioClient             : [3]: curr:1, closed:0,
org.tio.client.TioClient             : [4]: curr:1, closed:0,
org.tio.client.TioClient             : [5]: curr:1, closed:0,
org.tio.client.TioClient             : [2]: curr:0, closed:1,
```

图 4-10 运行过程中控制台显示所有有效连接(除了 2 号连接关闭之外其他连接有效)

```
出现异常: Address already in use: bind
```

图 4-11 服务器正常运行时再次在浏览器中输入 localhost:8080/tioserver 后控制台的输出结果

图 4-12 服务器正常运行时再次在浏览器中输入 localhost:8080/tioserver 后浏览器的输出结果

4.4 基于 ZooKeeper 实现 P2P 服务

视频讲解

　　Apache ZooKeeper(简称 ZooKeeper)是一种为分布式应用设计的服务协调框架,它提供配置管理、名字服务、分布式同步和集群管理等服务。ZooKeeper 有三种安装方式:单机模式、伪集群模式和集群模式。在单机模式下 ZooKeeper 只运行在一台服务器上,适合于开发、测试环境。伪集群模式是在一台物理机上运行多个 ZooKeeper 实例。在集群模式下,ZooKeeper 运行于一个集群上,适合于真实的生产环境。Spring Cloud ZooKeeper 为 Spring Cloud 应用提供 Apache ZooKeeper 的集成、封装和应用。本节以单机模式为例介绍如何基于 Spring Cloud ZooKeeper 实现 P2P 服务。

4.4.1 服务提供者模块 provider 的实现

　　参考附录 D 用 IDEA 创建多模块项目 zookeeperP2P,添加 provider 模块之后,确保在文件 pom.xml 中添加了 Web、Spring Cloud Zookeeper、Spring Cloud 等依赖,文件 pom.xml 的代码如例 4-19 所示。

【例 4-19】 文件 pom.xml 的代码示例。

```xml
<?xml version = "1.0" encoding = "UTF-8"?>
<project xmlns = "http://maven.apache.org/POM/4.0.0" xmlns:xsi = "http://www.w3.org/2001/XMLSchema-instance"
    xsi:schemaLocation = "http://maven.apache.org/POM/4.0.0 http://maven.apache.org/xsd/maven-4.0.0.xsd">
    <modelVersion>4.0.0</modelVersion>
    <parent>
        <groupId>org.springframework.boot</groupId>
        <artifactId>spring-boot-starter-parent</artifactId>
        <version>2.1.2.RELEASE</version>
```

```xml
        <relativePath/> <!-- lookup parent from repository -->
    </parent>
    <groupId>com.bookcode</groupId>
    <artifactId>provider</artifactId>
    <version>0.0.1-SNAPSHOT</version>
    <name>provider</name>
    <description>Demo project for Spring Boot</description>
    <properties>
        <java.version>1.8</java.version>
        <spring-cloud.version>Greenwich.RELEASE</spring-cloud.version>
    </properties>
    <dependencies>
        <dependency>
            <groupId>org.springframework.boot</groupId>
            <artifactId>spring-boot-starter-web</artifactId>
        </dependency>
        <dependency>
            <groupId>org.springframework.cloud</groupId>
            <artifactId>spring-cloud-starter-zookeeper-discovery</artifactId>
        </dependency>
        <dependency>
            <groupId>org.springframework.boot</groupId>
            <artifactId>spring-boot-starter-test</artifactId>
            <scope>test</scope>
        </dependency>
    </dependencies>
    <dependencyManagement>
        <dependencies>
            <dependency>
                <groupId>org.springframework.cloud</groupId>
                <artifactId>spring-cloud-dependencies</artifactId>
                <version>${spring-cloud.version}</version>
                <type>pom</type>
                <scope>import</scope>
            </dependency>
        </dependencies>
    </dependencyManagement>
    <build>
        <plugins>
            <plugin>
                <groupId>org.springframework.boot</groupId>
                <artifactId>spring-boot-maven-plugin</artifactId>
            </plugin>
        </plugins>
    </build>
    <repositories>
        <repository>
            <id>spring-milestones</id>
            <name>Spring Milestones</name>
            <url>https://repo.spring.io/milestone</url>
        </repository>
```

```
    </repositories>
</project>
```

在包 com.bookcode 中创建 controller 子包,并在包 com.bookcode.controller 中创建类 ProviderController,代码如例 4-20 所示。

【例 4-20】 类 ProviderController 的代码示例。

```
package com.bookcode.controller;
import org.springframework.beans.factory.annotation.Autowired;
import org.springframework.beans.factory.annotation.Value;
import org.springframework.web.bind.annotation.GetMapping;
import org.springframework.web.bind.annotation.PathVariable;
import org.springframework.web.bind.annotation.RestController;
import org.springframework.web.client.RestTemplate;
@RestController
public class ProviderController {
    @Value("${mypublicadress}")
    private String myadress;
    @Autowired
    RestTemplate restTemplate;
    @GetMapping("/receivefrom/{number}")
    public String sendBitCoin(@PathVariable(value = "number") int number ) {
return restTemplate.getForEntity( "http://consumer/getfrom/" + myadress + "/" + number,
String.class).getBody();
    }
    @GetMapping("/giveto/{name}/{number}")
    public String toBitCoin(@PathVariable(value = "name") String name,@PathVariable(value
= "number") int number ) {
        return "我发送给" + name + "比特币" + number + "元.";
    }
}
```

修改配置文件 application.properties,修改后的代码如例 4-21 所示。

【例 4-21】 修改后的配置文件 application.properties 的代码示例。

```
server.port = 8822
spring.application.name = spring-cloud-zookeeper-provider
spring.cloud.zookeeper.connectString = localhost:2181
mypublicadress = 3monidizhixinixinf
```

修改入口类,修改后的入口类代码如例 4-22 所示。

【例 4-22】 修改后的入口类代码示例。

```
package com.bookcode;
import org.springframework.boot.SpringApplication;
import org.springframework.boot.autoconfigure.SpringBootApplication;
import org.springframework.cloud.client.discovery.EnableDiscoveryClient;
```

```
import org.springframework.cloud.client.loadbalancer.LoadBalanced;
import org.springframework.context.annotation.Bean;
import org.springframework.web.client.RestTemplate;
@EnableDiscoveryClient
@SpringBootApplication
public class ProviderApplication {
    @Bean
    @LoadBalanced
    public RestTemplate restTemplate() {
        return new RestTemplate();
    }
    public static void main(String[] args) {
        SpringApplication.run(ProviderApplication.class, args);
    }
}
```

4.4.2 消费者模块 consumer 的实现

向项目添加消费者模块 consumer 之后,确保在文件 pom.xml 中添加了 Web、Spring Cloud ZooKeeper、Spring Cloud 等依赖,代码如例 4-19 所示。

在包 com.bookcode 中创建 controller 子包,并在包 com.bookcode.controller 中创建类 ConsumerController,代码如例 4-23 所示。

【例 4-23】 类 ConsumerController 的代码示例。

```
package com.bookcode.controller;
import org.springframework.beans.factory.annotation.Autowired;
import org.springframework.beans.factory.annotation.Value;
import org.springframework.web.bind.annotation.GetMapping;
import org.springframework.web.bind.annotation.PathVariable;
import org.springframework.web.bind.annotation.RestController;
import org.springframework.web.client.RestTemplate;
@RestController
public class ConsumerController {
    @Autowired
    RestTemplate restTemplate;
    @Value("${mypublicadress}")
    private String myadress;
    @GetMapping(value = "/getfrom/{name}/{number}")
    public String getBitCoin(@PathVariable(value = "name") String name,@PathVariable(value = "number") int number ) {
        return name + "发送比特币" + number + "元给我.";
    }
    @GetMapping(value = "/sendto/{number}")
    public String reciveBitCoin(@PathVariable(value = "number") int number ) {
        return restTemplate.getForEntity("http://provider/giveto/" + myadress + "/" + number, String.class).getBody();
```

```
    }
}
```

修改配置文件 application.properties，修改后的代码如例 4-24 所示。

【例 4-24】 修改后的配置文件 application.properties 的代码示例。

```
server.port = 8833
spring.application.name = spring-cloud-zookeeper-client
spring.cloud.zookeeper.connectString = localhost:2181
mypublicadress = 1monidizhixinixinf
```

修改入口类，修改后的入口类代码如例 4-25 所示。

【例 4-25】 修改后的入口类代码示例。

```java
package com.bookcode;
import org.springframework.boot.SpringApplication;
import org.springframework.boot.autoconfigure.SpringBootApplication;
import org.springframework.cloud.client.discovery.EnableDiscoveryClient;
import org.springframework.cloud.client.loadbalancer.LoadBalanced;
import org.springframework.context.annotation.Bean;
import org.springframework.web.client.RestTemplate;
@EnableDiscoveryClient
@SpringBootApplication
public class ConsumerApplication {
    @Bean
    @LoadBalanced
    public RestTemplate restTemplate() {
        return new RestTemplate();
    }
    public static void main(String[] args) {
        SpringApplication.run(ConsumerApplication.class, args);
    }
}
```

4.4.3 运行程序

参考附录 E 安装配置好 ZooKeeper 后，依次双击 zookeeper\bin 目录（如 D:\zookeeper\bin，和用户的 ZooKeeper 安装、配置目录有关）下文件 zkServer.cmd 和文件 zkCli.cmd，启动 ZooKeeper 的服务器和客户端。

依次运行模块 provider 的 ProviderApplication 和模块 consumer 的 ConsumerApplication。

在浏览器中输入 localhost:8822/receivefrom/1，浏览器输出结果如图 4-13 所示。在浏览器中输入 localhost:8822/giveto/zs/1，浏览器输出结果如图 4-14 所示。

在浏览器中输入 localhost:8833/sendto/1，浏览器输出结果如图 4-15 所示。在浏览器中输入 localhost:8833/getfrom/zs/1，浏览器输出结果如图 4-16 所示。

图 4-13　在浏览器中输入 localhost：8822/receivefrom/1 后浏览器输出的结果

图 4-14　在浏览器中输入 localhost：8822/giveto/zs/1 后浏览器输出的结果

图 4-15　在浏览器中输入 localhost：8833/sendto/1 后浏览器输出的结果

图 4-16　在浏览器中输入 localhost：8833/getfrom/zs/1 后浏览器输出的结果

4.5　基于 Web Service 和 CXF 实现 P2P 服务

视频讲解

本节介绍如何基于 Web Service 和 CXF 实现 P2P 服务。其中，基于 Web Service 实现服务器端，基于 CXF 实现客户端。

4.5.1　服务器端模块 serverofws 的实现

参考附录 D 用 IDEA 创建多模块项目 webservicep2p，添加 serverofws 模块之后，确保在文件 pom.xml 中<dependencies>和</dependencies>之间添加 Web、Web Services 依赖，代码如例 4-26 所示。

【例 4-26】　添加 Web、Web Services 依赖的代码示例。

```
<dependency>
          <groupId>org.springframework.boot</groupId>
          <artifactId>spring-boot-starter-web</artifactId>
</dependency>
```

```xml
<dependency>
    <groupId>org.springframework.boot</groupId>
    <artifactId>spring-boot-starter-web-services</artifactId>
</dependency>
```

在包 com.bookcode 中创建 service 子包,并在包 com.bookcode.service 中创建接口 IWebService,代码如例 4-27 所示。

【例 4-27】 接口 IWebService 的代码示例。

```java
package com.bookcode.service;
import javax.jws.WebMethod;
@javax.jws.WebService
public interface IWebService {
    @WebMethod
    String sayHello(String name);
}
```

在包 com.bookcode.service 中创建子包 impl,并在包 com.bookcode.service.impl 中创建类 WebServiceImpl,代码如例 4-28 所示。

【例 4-28】 类 WebServiceImpl 的代码示例。

```java
package com.bookcode.service.impl;
import com.bookcode.service.IWebService;
import java.util.Date;
@javax.jws.WebService
public class WebServiceImpl implements IWebService {
    @Override
    public String sayHello(String name) {
        String said = "Hello～～," + name + ",现在时间: " + "(" + new Date() + ")";
        System.out.println(said);
        return said;
    }
}
```

在包 com.bookcode 中创建 controller 子包,并在包 com.bookcode.controller 中创建类 PublishController,代码如例 4-29 所示。

【例 4-29】 类 PublishController 的代码示例。

```java
package com.bookcode.controller;
import com.bookcode.service.impl.WebServiceImpl;
import org.springframework.web.bind.annotation.GetMapping;
import org.springframework.web.bind.annotation.RestController;
import javax.xml.ws.Endpoint;
@RestController
public class PublishController {
    @GetMapping("/PeerofServer")
    public String serverrun(){
```

```
            String url = "http://localhost:9000/wsServeice";
            Endpoint.publish(url,new WebServiceImpl());
            return "发布 WebService 成功!";
    }
}
```

修改配置文件 application.properties,修改后的代码如例 4-30 所示。

【例 4-30】 修改后的配置文件 application.properties 代码示例。

```
server.port = 7000
```

4.5.2 客户端模块 clientofws 的实现

向项目添加客户端模块 clientofws 之后,确保在文件 pom.xml 中添加了 Web、CXF 相关依赖,代码如例 4-31 所示。

【例 4-31】 添加 Web、CXF 相关依赖的代码示例。

```xml
<dependency>
            <groupId>org.springframework.boot</groupId>
            <artifactId>spring-boot-starter-web</artifactId>
</dependency>
<dependency>
            <groupId>org.apache.cxf</groupId>
            <artifactId>cxf-rt-transports-http</artifactId>
            <version>3.2.5</version>
</dependency>
<dependency>
            <groupId>org.apache.cxf</groupId>
            <artifactId>cxf-rt-frontend-jaxws</artifactId>
            <version>3.1.12</version>
</dependency>
```

在包 com.bookcode 中创建 controller 子包,并在包 com.bookcode.controller 中创建类 CallController,代码如例 4-32 所示。

【例 4-32】 类 CallController 的代码示例。

```java
package com.bookcode.controller;
import org.springframework.web.bind.annotation.GetMapping;
import org.springframework.web.bind.annotation.RestController;
import org.apache.cxf.endpoint.Client;
import org.apache.cxf.jaxws.endpoint.dynamic.JaxWsDynamicClientFactory;
import org.apache.cxf.transport.http.HTTPConduit;
import org.apache.cxf.transports.http.configuration.HTTPClientPolicy;
```

```
@RestController
public class CallController {
    @GetMapping("/PeerofClient")
    private String clientrun(){
        String result = " Client OK !";
JaxWsDynamicClientFactory factory = JaxWsDynamicClientFactory.newInstance(); //动态客户端
        Client client = factory.createClient("http://localhost:9000/wsServeice?wsdl");
        HTTPConduit conduit = (HTTPConduit) client.getConduit();
        HTTPClientPolicy httpClientPolicy = new HTTPClientPolicy();
        httpClientPolicy.setConnectionTimeout(2000);      //连接超时
        httpClientPolicy.setAllowChunking(false);         //取消块编码
        httpClientPolicy.setReceiveTimeout(120000);       //响应超时
        conduit.setClient(httpClientPolicy);
        try{
            Object[] objects = new Object[0];
            objects = client.invoke("sayHello", "lisi");
            result = "返回数据:" + objects[0];
        }catch (Exception e){
            e.printStackTrace();
        }
        return result;
    }
}
```

4.5.3 运行程序

运行服务器端模块 serverofws 的 ServerofwsApplication 后，在浏览器中输入 localhost:7000/PeerofServer，浏览器输出结果如图 4-17 所示。再运行客户端模块 clientofws 的 ClientofwsApplication，在浏览器中输入 localhost:8080/PeerofClient，浏览器输出结果如图 4-18 所示。

图 4-17 在浏览器中输入 localhost:7000/PeerofServer 后浏览器输出的结果

图 4-18 在浏览器中输入 localhost:8080/PeerofClient 后浏览器输出的结果

4.6 同一服务器向多个页面发送区块链信息的示例

区块链应用往往需要将最新的区块链信息进行共享。本节通过同一个服务器向多个页面发送区块链信息来模拟区块链信息的共享。

4.6.1 创建项目并添加依赖

视频讲解

用 IDEA 创建完项目 exblock2p 之后,确保在文件 pom.xml 的< dependencies >和</dependencies >之间添加了 Web、Netty-socketio、Lombok、Fastjson 依赖,代码如例 3-33 所示。

【例 4-33】 添加 Web、Netty-socketio、Lombok、Fastjson 依赖的代码示例。

```
< dependency >
            < groupId > org.springframework.boot </groupId >
            < artifactId > spring - boot - starter - web </artifactId >
</dependency >
< dependency >
            < groupId > com.corundumstudio.socketio </groupId >
            < artifactId > netty - socketio </artifactId >
            < version > 1.7.16 </version >
</dependency >
< dependency >
            < groupId > org.projectlombok </groupId >
            < artifactId > lombok </artifactId >
            < version > 1.18.8 </version >
</dependency >
< dependency >
            < groupId > com.alibaba </groupId >
            < artifactId > fastjson </artifactId >
            < version > 1.2.54 </version >
</dependency >
```

4.6.2 创建类 Block

在包 com.bookcode 中创建 entity 子包,并在包 com.bookcode.entity 中创建类 Block,代码如例 4-34 所示。

【例 4-34】 类 Block 的代码示例。

```
package com.bookcode.entity;
import com.bookcode.utils.StringUtil;
import lombok.Data;
import java.util.Date;
@Data
public class Block {
```

```java
    private String hash;
    private String previousHash;
    private String data;
    private long timestamp;
    private int nonce;
    public Block(String data, String previousHash) {
        super();
        this.previousHash = previousHash;
        this.data = data;
        this.timestamp = new Date().getTime();
        this.hash = calculateHash();
    }
    public String calculateHash() {
      String calculatedHash = StringUtil.applySHA256 ( previousHash + Long.toString(timestamp) + data);
        return calculatedHash;

    }
}
```

4.6.3 创建类 SocketIOConfig

在包 com.bookcode 中创建 config 子包，并在包 com.bookcode.config 中创建类 SocketIOConfig，代码如例 3-35 所示。

【例 4-35】 类 SocketIOConfig 的代码示例。

```java
package com.bookcode.config;
import com.corundumstudio.socketio.SocketConfig;
import com.corundumstudio.socketio.SocketIOServer;
import org.springframework.beans.factory.annotation.Value;
import org.springframework.context.annotation.Bean;
import org.springframework.context.annotation.Configuration;
@Configuration
public class SocketIOConfig {
    @Value("${socketio.host}")
    private String host;
    @Value("${socketio.port}")
    private Integer port;
    @Value("${socketio.bossCount}")
    private int bossCount;
    @Value("${socketio.workCount}")
    private int workCount;
    @Value("${socketio.allowCustomRequests}")
    private boolean allowCustomRequests;
    @Value("${socketio.upgradeTimeout}")
    private int upgradeTimeout;
```

```java
    @Value("${socketio.pingTimeout}")
    private int pingTimeout;
    @Value("${socketio.pingInterval}")
    private int pingInterval;
    @Bean
    public SocketIOServer socketIOServer() {
        SocketConfig socketConfig = new SocketConfig();
        socketConfig.setTcpNoDelay(true);
        socketConfig.setSoLinger(0);
com.corundumstudio.socketio.Configuration config = new com.corundumstudio.socketio.Configuration();
        config.setSocketConfig(socketConfig);
        config.setHostname(host);
        config.setPort(port);
        config.setBossThreads(bossCount);
        config.setWorkerThreads(workCount);
        config.setAllowCustomRequests(allowCustomRequests);
        config.setUpgradeTimeout(upgradeTimeout);
        config.setPingTimeout(pingTimeout);
        config.setPingInterval(pingInterval);
        return new SocketIOServer(config);
    }
}
```

4.6.4 创建类 Service

在包 com.bookcode 中创建 common 子包,并在包 com.bookcode.common 中创建类 Service,代码如例 3-36 所示。

【例 4-36】 类 Service 的代码示例。

```java
package com.bookcode.common;
import com.alibaba.fastjson.JSONObject;
import com.bookcode.handler.MsgEventHandler;
import com.bookcode.entity.Block;
import com.bookcode.utils.LoggerUtil;
import lombok.Getter;
import lombok.Setter;
public class Service {
    @Getter
    @Setter
    private static String msg;
    static final Block[] fBlocks = new Block[20];
    //向"channel_1" 压入数据
    public static void send(String[] args) throws Exception {
        int price = 0;
        if (args != null && args.length > 0) {
            try {
```

```java
                price = Integer.parseInt(args[0]);
            } catch (Exception e) {
                LoggerUtil.log.info("args[0]不能转换为 int");
                new Exception(e);
            }
        }
        for (int i = 0; i < 6; i++) {
            JSONObject data = new JSONObject();
            createBlock(i * 3 + 1, i * 3 + 3);
            String str = "";
            for(int k = i * 3 + 1; k <= i * 3 + 3; k++){
                str += fBlocks[k].toString();
            }
            data.put("current_price", str);
            Service.msg = data.toJSONString(); //把每次压入的数据保存起来
            MsgEventHandler.sendAllEvent(data.toJSONString());
            Thread.sleep(5000 );
        }
    }
    private static void createBlock(int i, int j) {
        for(int k = i;k <= j; k++) {
            if (1 == k) {
                fBlocks[k] = new Block("我是第" + k + "个区块", "0");
            } else {
                fBlocks[k] = new Block("我是第" + k + "个区块", fBlocks[k - 1].calculateHash());
            }
        }
    }
}
```

4.6.5 创建类 ServerRunner

在包 com.bookcode.common 中创建类 ServerRunner,代码如例 3-37 所示。

【例 4-37】 类 ServerRunner 的代码示例。

```java
package com.bookcode.common;
import com.corundumstudio.socketio.SocketIOServer;
import com.corundumstudio.socketio.annotation.SpringAnnotationScanner;
import org.springframework.boot.CommandLineRunner;
import org.springframework.context.annotation.Bean;
import org.springframework.core.annotation.Order;
import org.springframework.stereotype.Component;
import javax.annotation.Resource;
@Component
@Order(1)
public class ServerRunner implements CommandLineRunner {
```

```
    @Resource
    SocketIOServer server;
    @Bean
    public SpringAnnotationScanner springAnnotationScanner(SocketIOServer socketServer) {
        return new SpringAnnotationScanner(socketServer);
    }
    @Override
    public void run(String... args) throws Exception {
        System.out.println("ServerRunner 开始启动啦...");
        server.start();
        Service.send(args);
    }
}
```

4.6.6 创建类 MsgEventHandler

在包 com.bookcode 中创建 handler 子包,并在包 com.bookcode.handler 中创建类 MsgEventHandler,代码如例 3-38 所示。

【例 4-38】 类 MsgEventHandler 的代码示例。

```
package com.bookcode.handler;
import com.bookcode.common.Service;
import com.bookcode.utils.LoggerUtil;
import com.corundumstudio.socketio.AckRequest;
import com.corundumstudio.socketio.SocketIOClient;
import com.corundumstudio.socketio.SocketIOServer;
import com.corundumstudio.socketio.annotation.OnConnect;
import com.corundumstudio.socketio.annotation.OnDisconnect;
import com.corundumstudio.socketio.annotation.OnEvent;
import org.springframework.beans.factory.annotation.Autowired;
import org.springframework.stereotype.Component;
import java.util.Set;
import java.util.UUID;
@Component
public class MsgEventHandler {
    public static SocketIOServer socketIoServer;
    @Autowired
    public MsgEventHandler(SocketIOServer server) {
        socketIoServer = server;
    }
    //在通道"channel_1"中广播消息
    public static void sendAllEvent(String data) {
        socketIoServer.getRoomOperations("channel_1").sendEvent("channel_1", data);
    }
    @OnConnect
    public void onConnect(SocketIOClient client) {
        UUID socketSessionId = client.getSessionId();
```

```java
        String ip = client.getRemoteAddress().toString();
        LoggerUtil.log.info("client connect, socketSessionId:{}, ip:{}", socketSessionId, ip);
    }
    @OnEvent("sub")
    public void sub(SocketIOClient client, AckRequest request, String channel) {
        UUID socketSessionId = client.getSessionId();
        String ip = client.getRemoteAddress().toString();
        client.joinRoom(channel);
        LoggerUtil.log.info("client sub, channel:{}, socketSessionId:{}, ip:{}", channel, socketSessionId, ip);
        Set<String> rooms = client.getAllRooms();
        for (String room : rooms) {
            LoggerUtil.log.info("after client connect, room:{}", room);
        }
        //客户端一订阅,就马上压入一次
        sendAllEvent(Service.getMsg());
    }
    @OnDisconnect
    public void onDisconnect(SocketIOClient client) {
        UUID socketSessionId = client.getSessionId();
        String ip = client.getRemoteAddress().toString();
        LoggerUtil.log.info("client disconnect, socketSessionId:{}, ip:{}", socketSessionId, ip);
        Set<String> rooms = client.getAllRooms();
        for (String room : rooms) {
            LoggerUtil.log.info("after client disconnect, room:{}", room);
        }
    }
}
```

4.6.7 创建类 ClientController

在包 com.bookcode 中创建 controller 子包,并在包 com.bookcode.controller 中创建类 ClientController,代码如例 4-39 所示。

【例 4-39】 类 ClientController 的代码示例。

```java
package com.bookcode.controller;
import org.springframework.stereotype.Controller;
import org.springframework.web.bind.annotation.GetMapping;
@Controller
public class ClientController {
    @GetMapping("/")
    public String testHome(){
        return "redirect:/index.html";
    }
}
```

```java
@GetMapping("/index")
public String testIndex(){
    return "redirect:/index.html";
}
@GetMapping("/welcome")
public String testWelcome(){
    return "redirect:/welcome.html";
}
@GetMapping("/login")
public String testLogin(){
    return "redirect:/login.html";
}
}
```

4.6.8 创建类 StringUtil

在包 com.bookcode 中创建 utils 子包,并在包 com.bookcode.utils 中创建类 StringUtil,代码如例 4-40 所示。

【例 4-40】 类 StringUtil 的代码示例。

```java
package com.bookcode.utils;
import java.io.UnsupportedEncodingException;
import java.security.MessageDigest;
import java.security.NoSuchAlgorithmException;
public class StringUtil {
    public static String applySHA256(String input) {
        StringBuffer hexString = null;
        try {
            MessageDigest messageDigest = MessageDigest.getInstance("SHA-256");
            byte[] hash = messageDigest.digest(input.getBytes("UTF-8"));
            hexString = new StringBuffer();
            String hex = null;
            for(int t = 0 ; t < hash.length ; t++) {
                hex = Integer.toHexString(0xff & hash[t]);
                if( hex.length() == 1 ) {
                    hexString.append('0');
                } else {
                    hexString.append(hex);
                }
            }
        } catch (NoSuchAlgorithmException e) {
            e.printStackTrace();
        } catch (UnsupportedEncodingException e) {
            e.printStackTrace();
        }
        return hexString.toString();
    }
}
```

4.6.9 创建类 LoggerUtil

在包 com.bookcode.utils 中创建类 LoggerUtil,代码如例 4-41 所示。

【例 4-41】 类 LoggerUtil 的代码示例。

```
package com.bookcode.utils;
import com.bookcode.common.ServerRunner;
import org.slf4j.Logger;
import org.slf4j.LoggerFactory;
public class LoggerUtil {
    public static Logger log = LoggerFactory.getLogger(ServerRunner.class);
}
```

4.6.10 创建文件 index.html、login.html 和 welcome.html

在目录 src/main/resources/static 下,创建文件 index.html、login.html 和 welcome.html,三个文件的代码完全相同(只是文件名不同),如例 4-42 所示。

【例 4-42】 文件 index.html、login.html 和 welcome.html 的代码示例。

```html
<!DOCTYPE html PUBLIC "-//W3C//DTD HTML 4.01 Transitional//EN" "http://www.w3.org/TR/html4/loose.dtd">
<html>
<head>
    <meta http-equiv="Content-Type" content="text/html; charset=UTF-8">
    <title>Insert title here</title>
    <script src="http://code.jquery.com/jquery-1.12.4.min.js"></script>
    <script type="text/javascript" src="https://cdn.bootcss.com/socket.io/2.2.0/socket.io.js"></script>
    <style>
        body {
            padding:20px;
        }
        #console {
            overflow: auto;
        }
        .username-msg {color:orange;}
        .connect-msg {color:green;}
        .disconnect-msg {color:red;}
        .send-msg {color:#888}
    </style>
</head>
<body>
<div id="console" class="well">
</div><br/><br/>
收到的最新 3 个区块链信息:      <br>
```

```html
<span id="current_price"></span><br/>
</body>
<script type="text/javascript">
    var socket = io.connect('http://localhost:9099', {
        transports:['websocket']
    });
    socket.on('connect', function() {
        console.log("msg页面连接成功!");
        socket.emit('sub', "channel_1");
        output('<span class="connect-msg">Client has connected to the server!</span>');
   output('<span class="connect-msg">Client send {"event": "sub", "channel": "channel_1"}</span>');
    });
    socket.on('disconnect', function() {
        output('<span class="disconnect-msg">The client has disconnected!</span>');
    });
    socket.on('channel_1', function(data) {
        var jsonObj = eval("(" + data + ")");
        console.log("收到cfd_md的消息:" + data);
         $("#current_price").html(jsonObj.current_price);
    });
    function output(message) {
        var currentTime = "<span class='time'>" + NowTime() + "</span>";
        var element = $("<div>" + currentTime + " " + message + "</div>");
         $('#console').prepend(element);
    }
    //获取当前时间
    function NowTime() {
        var time = new Date();
        var year = time.getFullYear();        //获取年
        var month = time.getMonth() + 1;      //获取月
        var day = time.getDate();             //获取天
        var hour = time.getHours();           //获取小时
        var minu  = time.getMinutes();        //获取分钟
        var second = time.getSeconds();       //获取秒
        var data = year + "-";
        if(month < 10){
            data += "0";
        }
        data += month + "-";
        if(day < 10){
            data += "0"
        }
        data += day + " ";
        if(hour < 10){
            data += "0"
        }
        data += hour + ":";
        if(minu < 10){
            data += "0"
```

```
            }
            data += minu + ":";
            if(second < 10){
                data += "0"
            }
            data += second;
            return data;
        }
    </script>
</html>
```

4.6.11 修改配置文件 application.properties

修改配置文件 application.properties，修改后的代码如例 4-43 所示。

【例 4-43】 修改后的配置文件 application.properties 代码示例。

```
#socketio 设置信息
#host 在本地测试可以设置为 localhost 或者本机 IP,在 Linux 服务器可换成服务器 IP
socketio.host = localhost
socketio.port = 9099
#设置每帧处理的最大数据长度,防止他人利用大数据来攻击服务器
socketio.maxFramePayloadLength = 1048576
#设置 HTTP 交互最大内容长度
socketio.maxHttpContentLength = 1048576
#socket 连接数大小(如只监听一个端口 boss 线程组为 1 即可)
socketio.bossCount = 1
socketio.workCount = 100
socketio.allowCustomRequests = true
#协议升级超时时间(毫秒),默认为 10s。HTTP 握手升级为 ws 协议超时时间
socketio.upgradeTimeout = 1000000
#Ping 消息超时时间(毫秒),默认为 60s,这个时间间隔内没有接收到心跳消息就会发送超时事件
socketio.pingTimeout = 6000000
#Ping 消息间隔(毫秒),默认为 25s。客户端向服务器发送一条心跳消息间隔
socketio.pingInterval = 25000
```

4.6.12 运行程序

运行程序后,在浏览器中输入 localhost:8080/index.html,控制台中输出结果显示服务器正常启动,如图 4-19 所示。接着,控制台输出服务器和第一个页面的连接信息(端口为 61523),如图 4-20 所示。请注意,端口是动态分配的,每次运行程序可能不同。浏览器的最终输出结果(第 16、17、18 个区块)如图 4-21 所示。请注意,浏览器中输出的区块链信息动态改变,从第 1 个区块一直到第 18 区块,每 3 个区块为一组(即从 1、2、3 到 16、17、18)。

接着,在浏览器中输入 localhost:8080/index.html,localhost:8080/login.html 或者 localhost:8080/welcome.html 等(输出结果和在浏览器中输入的 URL 顺序无关),控制台

会输出服务器和新的页面连接信息（如端口 61538、61554、61557 等），控制台输出结果如图 4-22 所示。与此同时，浏览器中输出结果与图 4-21 相同（URL 内容有差异）。

ServerRunner 开始启动啦...

图 4-19　控制台中输出结果显示服务器正常启动

```
client connect, socketSessionId:3d2e2fb4-2e8e-43ff-a251-47f5a7053f42, ip:/127.0.0.1:61523
client sub, channel:channel_1, socketSessionId:3d2e2fb4-2e8e-43ff-a251-47f5a7053f42, ip:/127.0.0.1:61523
after client connect, room:
after client connect, room:channel_1
client connect, socketSessionId:bbaa3602-6089-4627-9fc7-d03c95660ed5, ip:/127.0.0.1:61538
client sub, channel:channel_1, socketSessionId:bbaa3602-6089-4627-9fc7-d03c95660ed5, ip:/127.0.0.1:61538
after client connect, room:
after client connect, room:channel_1
```

图 4-20　客户端页面打开后控制台中输出服务器和客户端的连接信息

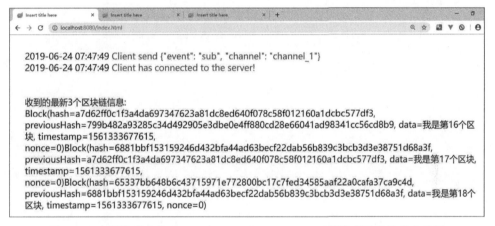

图 4-21　在浏览器中输入 localhost:8080/index.html 后浏览器的最终输出结果

```
after client connect, room:
after client connect, room:channel_1
client disconnect, socketSessionId:bbaa3602-6089-4627-9fc7-d03c95660ed5, ip:/127.0.0.1:61538
client connect, socketSessionId:9b7ea534-2e74-4bd6-bb02-393344b2da2a, ip:/127.0.0.1:61546
client sub, channel:channel_1, socketSessionId:9b7ea534-2e74-4bd6-bb02-393344b2da2a, ip:/127.0.0.1:61546
after client connect, room:
after client connect, room:channel_1
client connect, socketSessionId:1dc1fed4-47d8-4e7f-a12a-193a5d4c84f7, ip:/127.0.0.1:61554
client sub, channel:channel_1, socketSessionId:1dc1fed4-47d8-4e7f-a12a-193a5d4c84f7, ip:/127.0.0.1:61554
after client connect, room:
after client connect, room:channel_1
client disconnect, socketSessionId:1dc1fed4-47d8-4e7f-a12a-193a5d4c84f7, ip:/127.0.0.1:61554
client connect, socketSessionId:704ebf37-4e73-4a59-9cd0-2caffba53611, ip:/127.0.0.1:61557
client sub, channel:channel_1, socketSessionId:704ebf37-4e73-4a59-9cd0-2caffba53611, ip:/127.0.0.1:61557
after client connect, room:
after client connect, room:channel_1
```

图 4-22　多个客户端页面打开后控制台中输出服务器和客户端的连接信息

习题 4

实验题
1. 基于 Java-WebSocket 实现 P2P 网络。
2. 基于 WebSocket 实现 P2P 页面互连。
3. 基于 t-io 实现 P2P 网络。
4. 基于 ZooKeeper 实现 P2P 服务。
5. 基于 Web Service 和 CXF 实现 P2P 服务。

第5章

区块链应用中消息机制的实现

JMS(Java Message Service,Java 消息服务)API 是消息服务的标准、规范,它允许应用程序组件基于 Java EE 平台创建、发送、接收和读取消息。它使分布式通信耦合度更低、消息服务更加可靠,更好地实现异步性。

本章依次介绍 ActiveMQ 的应用、RabbitMQ 的应用、Spring Cloud Stream 和 RabbitMQ 的应用以及基于 ActiveMQ 传递区块链消息的示例。

5.1 ActiveMQ 的应用

ActiveMQ 是由 Apache 出品的一款流行的开源消息总线。ActiveMQ 是一个完全支持 JMS 和 Java EE 规范的 JMS Provider 实现,它运行速度快,支持多种语言的客户端和协议,而且可以很容易地嵌入企业的应用环境中。

视频讲解

5.1.1 创建项目并添加依赖

用 IDEA 创建完项目 amqexample 之后,确保在文件 pom.xml 中< dependencies >和</dependencies >之间添加了 Web、JMS、ActiveMQ 相关依赖,代码如例 5-1 所示。

【例 5-1】 添加 Web、JMS、ActiveMQ 相关依赖的代码示例。

```
<dependency>
    <groupId>org.springframework.boot</groupId>
    <artifactId>spring-boot-starter-web</artifactId>
</dependency>
<dependency>
    <groupId>javax.jms</groupId>
    <artifactId>javax.jms-api</artifactId>
```

```xml
</dependency>
<dependency>
        <groupId>org.apache.activemq</groupId>
        <artifactId>activemq-core</artifactId>
        <version>5.7.0</version>
</dependency>
<dependency>
        <groupId>org.apache.activemq</groupId>
        <artifactId>activemq-all</artifactId>
        <version>5.15.8</version>
</dependency>
```

5.1.2 创建类 Sender

在包 com.bookcode 中创建 service 子包,并在包 com.bookcode.service 中创建类 Sender,代码如例 5-2 所示。

【例 5-2】 类 Sender 的代码示例。

```java
package com.bookcode.service;
import javax.jms.Connection;
import javax.jms.ConnectionFactory;
import javax.jms.Destination;
import javax.jms.JMSException;
import javax.jms.MessageProducer;
import javax.jms.Session;
import javax.jms.TextMessage;
import com.bookcode.util.RsUtil;
import org.apache.activemq.ActiveMQConnectionFactory;
import org.springframework.stereotype.Service;
@Service
public class Sender {
    public String initSender() {
        ConnectionFactory connectionFactory;
        Connection connection = null;
        Session session;
        Destination destination;
        MessageProducer messageProducer;
        connectionFactory = new ActiveMQConnectionFactory(RsUtil.USERNAME, RsUtil.PASSWORD, RsUtil.BROKEURL);
        try {
            connection = connectionFactory.createConnection();
            connection.start();
            session = connection.createSession(true, Session.AUTO_ACKNOWLEDGE);
            destination = session.createQueue("HelloWorld");
            messageProducer = session.createProducer(destination);
            sendMessage(session, messageProducer);
            session.commit();
```

```java
            } catch (Exception e) {
                e.printStackTrace();
            }finally{
                if(connection != null){
                    try {
                        connection.close();
                    } catch (JMSException e) {
                        e.printStackTrace();
                    }
                }
            }
        return "Send OK";
    }
    private void sendMessage ( Session session, MessageProducer messageProducer ) throws Exception{
        for ( int i = 0; i < RsUtil.SENDNUM; i++) {
            TextMessage message = session.createTextMessage("ActiveMQ 发送消息" + i);
            System.out.println("发送消息: Activemq 发送消息" + i);
            messageProducer.send(message);
        }
    }
}
```

5.1.3 创建类 Receiver

在包 com.bookcode.service 中创建类 Receiver，代码如例 5-3 所示。

【例 5-3】 类 Receiver 的代码示例。

```java
package com.bookcode.service;
import com.bookcode.util.RsUtil;
import javax.jms.Connection;
import javax.jms.ConnectionFactory;
import javax.jms.Destination;
import javax.jms.JMSException;
import javax.jms.MessageConsumer;
import javax.jms.Session;
import javax.jms.TextMessage;
import org.apache.activemq.ActiveMQConnectionFactory;
import org.springframework.stereotype.Service;
@Service
public class Receiver {
    public String initReceiver(){
        ConnectionFactory connectionFactory;
        Connection connection = null;
        Session session;
        Destination destination;
        MessageConsumer messageConsumer;
```

```
                    connectionFactory = new ActiveMQConnectionFactory(RsUtil.USERNAME, RsUtil.
PASSWORD,RsUtil.BROKEURL);
        try {
            connection = connectionFactory.createConnection();
            connection.start();
            session = connection.createSession(false, Session.AUTO_ACKNOWLEDGE);
            destination = session.createQueue("HelloWorld");
            messageConsumer = session.createConsumer(destination);
            while (true) {
                TextMessage textMessage = (TextMessage) messageConsumer.receive(100000);
                if(textMessage != null){
                    System.out.println("收到的消息:" + textMessage.getText());
                }else {
                    break;
                }
            }
        } catch (JMSException e) {
            e.printStackTrace();
        }
        return "Recive OK";
    }
}
```

5.1.4　创建类 RsUtil

在包 com.bookcode 中创建 util 子包，并在包 com.bookcode.util 中创建类 RsUtil，代码如例 5-4 所示。

【例 5-4】　类 RsUtil 的代码示例。

```
package com.bookcode.util;
import org.apache.activemq.ActiveMQConnection;
public class RsUtil {
    public static final String USERNAME = ActiveMQConnection.DEFAULT_USER;
    public static final String PASSWORD = ActiveMQConnection.DEFAULT_PASSWORD;
    public static final String BROKEURL = ActiveMQConnection.DEFAULT_BROKER_URL;
    public static final int SENDNUM = 10;
}
```

5.1.5　创建类 ActiveMQController

在包 com.bookcode 中创建 controller 子包，在包 com.bookcode.controller 中创建类 ActiveMQController，代码如例 5-5 所示。

【例 5-5】　类 ActiveMQController 的代码示例。

```
package com.bookcode.controller;
import com.bookcode.service.Receiver;
import com.bookcode.service.Sender;
import org.springframework.beans.factory.annotation.Autowired;
import org.springframework.web.bind.annotation.GetMapping;
import org.springframework.web.bind.annotation.RestController;
@RestController
public class ActiveMQController {
    @Autowired
    private Sender tioServer;
    @Autowired
    private Receiver tioClient;
    @GetMapping("/tioserver")
    public String server() {
        return tioServer.initSender();
    }
    @GetMapping("/tioclient")
    public String client() {
        return tioClient.initReceiver();
    }
}
```

5.1.6 运行程序

参考附录 F 配置好 ActiveMQ 后,双击 bin 目录(如 D:\apache-activemq-5.9.0\bin,和用户的目录有关)下文件 activemq.bat,启动 ActiveMQ。

运行程序后,在浏览器中输入 localhost:8080/tioserver,浏览器的输出结果如图 5-1 所示,控制台的输出结果如图 5-2 所示。在浏览器中输入 localhost:8080/tioclient,浏览器的输出结果如图 5-3 所示,控制台的输出结果如图 5-4 所示。

图 5-1 在浏览器中输入 localhost:8080/tioserver 后浏览器的输出结果

```
Successfully connected to tcp://localhost:61616
发送消息: Activemq 发送消息 0
发送消息: Activemq 发送消息 1
发送消息: Activemq 发送消息 2
发送消息: Activemq 发送消息 3
发送消息: Activemq 发送消息 4
发送消息: Activemq 发送消息 5
发送消息: Activemq 发送消息 6
发送消息: Activemq 发送消息 7
发送消息: Activemq 发送消息 8
发送消息: Activemq 发送消息 9
```

图 5-2 在浏览器中输入 localhost:8080/tioserver 后控制台的输出结果

图 5-3　在浏览器中输入 localhost:8080/tioclient 后浏览器的输出结果

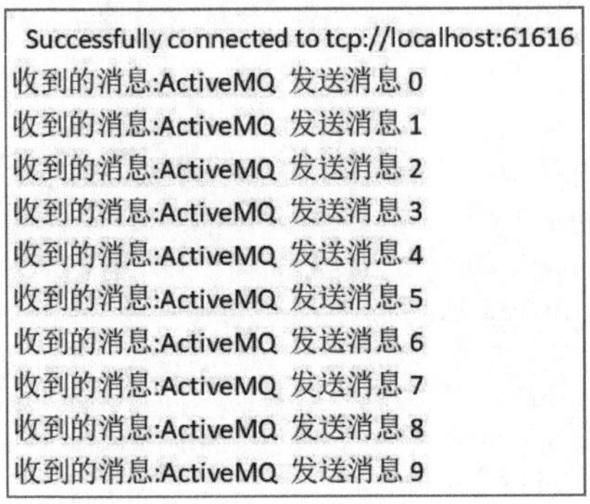

图 5-4　在浏览器中输入 localhost:8080/tioclient 后控制台的输出结果

5.2　RabbitMQ 的应用

视频讲解

RabbitMQ 是实现了高级消息队列协议（Advanced Message Queuing Protocol,AMQP）的开源消息队列服务软件（也称面向消息的中间件）。本节主要介绍 RabbitMQ 的应用。

5.2.1　创建项目并添加依赖

用 IDEA 创建完项目 cdbexample 之后，确保在文件 pom.xml 中添加了 Rabbit 依赖，代码如例 5-6 所示。

【例 5-6】　添加 Rabbit 依赖的代码示例。

```
<dependency>
        <groupId>org.springframework.amqp</groupId>
        <artifactId>spring-rabbit</artifactId>
        <version>2.1.4.RELEASE</version>
</dependency>
```

5.2.2　创建类 Runner

在包 com.auroralove 中创建类 Runner,代码如例 5-7 所示。

【例 5-7】　类 Runner 的代码示例。

```
package com.auroralove;
import java.util.concurrent.TimeUnit;
import org.springframework.amqp.rabbit.core.RabbitTemplate;
import org.springframework.boot.CommandLineRunner;
import org.springframework.stereotype.Component;
@Component
public class Runner implements CommandLineRunner {
    private final RabbitTemplate rabbitTemplate;
    private final Receiver receiver;
    public Runner(Receiver receiver, RabbitTemplate rabbitTemplate) {
        this.receiver = receiver;
        this.rabbitTemplate = rabbitTemplate;
    }
    @Override
    public void run(String... args) throws Exception {
        System.out.println("Sending message...");
rabbitTemplate.convertAndSend("spring-boot-exchange","foo.bar.baz","Hello from RabbitMQ!");
        receiver.getLatch().await(10000, TimeUnit.MILLISECONDS);
    }
}
```

5.2.3 创建类 Receiver

在包 com.auroralove 中创建类 Receiver，代码如例 5-8 所示。

【例 5-8】 类 Receiver 的代码示例。

```
package com.auroralove;
import java.util.concurrent.CountDownLatch;
import org.springframework.stereotype.Component;
@Component
public class Receiver {
    private CountDownLatch latch = new CountDownLatch(1);
    public void receiveMessage(String message) {
        System.out.println("Received <" + message + ">");
        latch.countDown();
    }
    public void receiveMessage(Object message) {
        System.out.println("Received <" + message + ">");
        latch.countDown();
    }
    public CountDownLatch getLatch() {
        return latch;
    }
}
```

5.2.4 修改入口类

修改入口类,修改后的入口类代码如例 5-9 所示。

【例 5-9】 修改后的入口类代码示例。

```java
package com.auroralove;
import org.springframework.amqp.core.Binding;
import org.springframework.amqp.core.BindingBuilder;
import org.springframework.amqp.core.Queue;
import org.springframework.amqp.core.TopicExchange;
import org.springframework.amqp.rabbit.connection.ConnectionFactory;
import org.springframework.amqp.rabbit.listener.SimpleMessageListenerContainer;
import org.springframework.amqp.rabbit.listener.adapter.MessageListenerAdapter;
import org.springframework.boot.SpringApplication;
import org.springframework.boot.autoconfigure.SpringBootApplication;
import org.springframework.context.annotation.Bean;
@SpringBootApplication
public class CdbexampleApplication {
    static final String queueName = "spring-boot-1";
    static final String queueName2 = "spring-boot-2";
    @Bean
    Queue queue() {
        return new Queue(queueName, false);
    }
    @Bean
    Queue queue2() {
        return new Queue(queueName2, false);
    }
    @Bean
    TopicExchange exchange() {
        return new TopicExchange("spring-boot-exchange");
    }
    @Bean
    Binding binding1(Queue queue, TopicExchange exchange) {
        return BindingBuilder.bind(queue).to(exchange).with("foo.bar.#");
    }
    @Bean
    Binding binding2(Queue queue2, TopicExchange exchange) {
        return BindingBuilder.bind(queue2).to(exchange).with("foo.#");
    }
    @Bean
    SimpleMessageListenerContainer container(ConnectionFactory connectionFactory,
                                    MessageListenerAdapter listenerAdapter) {
        SimpleMessageListenerContainer container = new SimpleMessageListenerContainer();
        container.setConnectionFactory(connectionFactory);
        container.setQueueNames(queueName);
        container.setMessageListener(listenerAdapter);
```

```
            return container;
        }
        @Bean
        MessageListenerAdapter listenerAdapter(Receiver receiver) {
            return new MessageListenerAdapter(receiver, "receiveMessage");
        }
        public static void main(String[] args) {
            SpringApplication.run(CdbexampleApplication.class, args);
        }
    }
```

5.2.5　运行程序

在 Windows 命令行程序中执行如例 5-10 所示的命令启动 RabbitMQ 服务。

【例 5-10】 启动 RabbitMQ 服务的命令示例。

```
rabbitmq-server
```

运行程序后,控制台的输出结果如图 5-5 所示。在浏览器中输入 localhost:15672/#/Queues,输出结果如图 5-6 所示。在浏览器中输入 localhost:15672/#/exchanges/%2F/spring-boot-exchange,输出结果如图 5-7 所示。

```
Sending message...
Received <Hello from RabbitMQ!>
```

图 5-5　控制台的输出结果

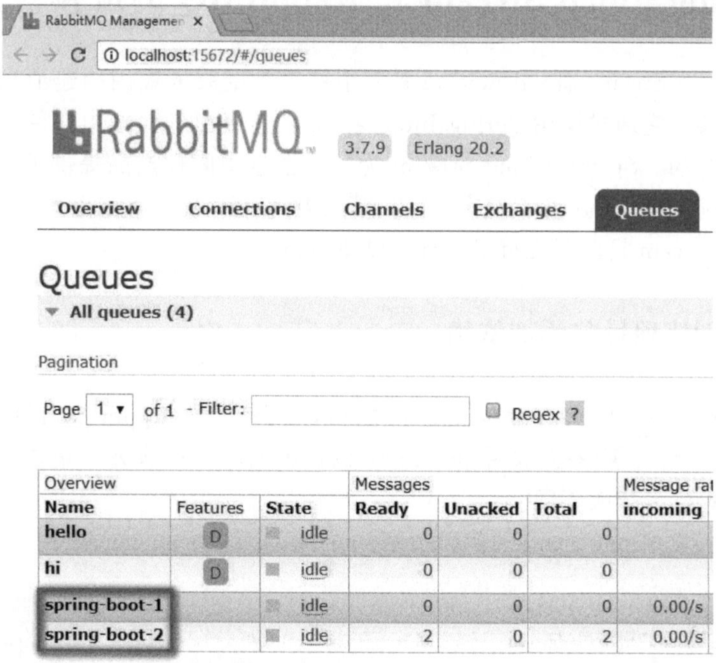

图 5-6　在浏览器中输入 localhost:15672/#/Queues 的输出结果

图 5-7　在浏览器中输入 localhost:15672/#/exchanges/%2F/spring-boot-exchange 的输出结果

5.3　Spring Cloud Stream 和 RabbitMQ 的应用

视频讲解

Spring Cloud Stream 是一个用来为微服务应用构建消息驱动能力的框架。它通过使用 Spring Integration 来连接消息代理中间件以实现消息事件驱动。Spring Cloud Stream 为一些消息中间件产品提供了个性化的自动化配置实现,引用了发布-订阅、消费组、分区的三个核心概念。Spring Cloud Stream 目前仅支持 RabbitMQ、Kafka。

5.3.1　创建项目并添加依赖

用 IDEA 创建完项目 streamexample 之后,确保在文件 pom.xml 中添加了 Spring Cloud、Stream Rabbit 等依赖,文件 pom.xml 的代码如例 5-11 所示。请注意,5.3 节和 5.2 节中对 RabbitMQ 的依赖不同。

【例 5-11】　文件 pom.xml 的代码示例。

```
<?xml version = "1.0" encoding = "UTF - 8"?>
< project xmlns = "http://maven.apache.org/POM/4.0.0" xmlns:xsi = "http://www.w3.org/2001/
XMLSchema - instance"
```

```xml
        xsi:schemaLocation = "http://maven.apache.org/POM/4.0.0 http://maven.apache.org/xsd/maven-4.0.0.xsd">
    <modelVersion>4.0.0</modelVersion>
    <groupId>com.bookcode</groupId>
    <artifactId>demo</artifactId>
    <version>0.0.1-SNAPSHOT</version>
    <packaging>jar</packaging>
    <name>demo</name>
    <description>Demo project for Spring Cloud</description>
    <parent>
        <groupId>org.springframework.boot</groupId>
        <artifactId>spring-boot-starter-parent</artifactId>
        <version>2.1.4.RELEASE</version>
        <relativePath/> <!-- lookup parent from repository -->
    </parent>
    <properties>
        <project.build.sourceEncoding>UTF-8</project.build.sourceEncoding>
        <project.reporting.outputEncoding>UTF-8</project.reporting.outputEncoding>
        <java.version>1.8</java.version>
        <spring-cloud.version>Greenwich.RELEASE</spring-cloud.version>
    </properties>
    <dependencies>
        <dependency>
            <groupId>org.springframework.cloud</groupId>
            <artifactId>spring-cloud-starter-stream-rabbit</artifactId>
        </dependency>
        <dependency>
            <groupId>org.springframework.cloud</groupId>
            <artifactId>spring-cloud-stream</artifactId>
            <version>2.1.0.RELEASE</version>
        </dependency>
        <dependency>
            <groupId>org.springframework.boot</groupId>
            <artifactId>spring-boot-starter-test</artifactId>
            <scope>test</scope>
        </dependency>
    </dependencies>
    <dependencyManagement>
        <dependencies>
            <dependency>
                <groupId>org.springframework.cloud</groupId>
                <artifactId>spring-cloud-dependencies</artifactId>
                <version>${spring-cloud.version}</version>
                <type>pom</type>
                <scope>import</scope>
            </dependency>
        </dependencies>
    </dependencyManagement>
    <build>
        <plugins>
```

```xml
            <plugin>
                <groupId>org.springframework.boot</groupId>
                <artifactId>spring-boot-maven-plugin</artifactId>
            </plugin>
        </plugins>
    </build>
    <repositories>
        <repository>
            <id>spring-milestones</id>
            <name>Spring Milestones</name>
            <url>https://repo.spring.io/milestone</url>
            <snapshots>
                <enabled>false</enabled>
            </snapshots>
        </repository>
    </repositories>
</project>
```

5.3.2 创建接口 Sink

在包 com.bookcode 中创建接口 Sink,代码如例 5-12 所示。

【例 5-12】 接口 Sink 的代码示例。

```java
package com.bookcode;
import org.springframework.cloud.stream.annotation.Input;
import org.springframework.messaging.SubscribableChannel;
public interface Sink {
    String INPUT = "input";
    @Input("input")
    SubscribableChannel input();
}
```

5.3.3 创建类 SinkReceiver

在包 com.bookcode 中创建类 SinkReceiver,代码如例 5-13 所示。

【例 5-13】 类 SinkReceiver 的代码示例。

```java
package com.bookcode;
import org.slf4j.Logger;
import org.slf4j.LoggerFactory;
import org.springframework.cloud.stream.annotation.EnableBinding;
import org.springframework.cloud.stream.annotation.StreamListener;
@EnableBinding(value = {Sink.class})    //用来指定定义了@Input 或者 @Output 注解的接口
//实现对消息通道的绑定.Sink 接口是默认输入消息通道绑定接口
```

```
public class SinkReceiver {
    private static Logger logger = LoggerFactory.getLogger(SinkReceiver.class);
    @StreamListener(Sink.INPUT)    //将被修饰的方法注册为消息中间件上数据流的事件监听器
    public void receive(String payload) {
         logger.info("SinkReceiver --- 接收到的信息: " + payload);
    }
}
```

5.3.4　创建配置文件 application.yml

在目录 src/main/resources 下创建配置文件 application.yml 并修改配置文件代码,修改后的代码如例 5-14 所示。

【例 5-14】　创建的配置文件 application.yml 代码示例。

```
logging:
    file: log\outputinfo.log
```

5.3.5　运行程序

在 Windows 命令行程序中执行如例 5-10 所示的命令启动 RabbitMQ 服务。

运行程序后,在浏览器中输入 localhost:15672/#/Queues,并在 Payload 后面的文本框中输入"您好,这是 RabbitMQ 发布的信息!",单击 Publish message 按钮,弹出提示信息框(内容为 Message published),结果如图 5-8 所示。完成上述操作后,控制台中的相关输出结果如图 5-9 所示。与此同时,根据配置文件的设置在项目中自动创建了一个子目录(\log),并在子目录中生成了一个日志文件(outputinfo.log),日志文件内容中的相关输出结果如图 5-10 所示。

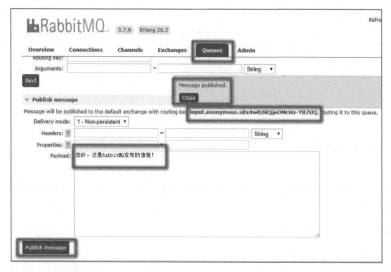

图 5-8　在浏览器中输入 localhost:15672/#/Queues 并进行相关操作的结果

```
com.bookcode.SinkReceiver                  : SinkReceiver---接收到的信息：您好！这是RabbitMQ发布的信息！
```

图 5-9 完成相关操作后控制台中的相关输出结果

```
com.bookcode.SinkReceiver                  : SinkReceiver---接收到的信息：您好！这是RabbitMQ发布的信息！
```

图 5-10 完成相关操作后日志文件内容中的相关输出结果

5.4 基于 ActiveMQ 传递区块链消息的示例

本节介绍如何用 ActiveMQ 来传递区块链消息。

5.4.1 创建项目并添加依赖

用 IDEA 创建完项目 examqbbc 之后，确保在文件 pom.xml 中<dependencies>和</dependencies>之间添加了 Web、ActiveMQ、Artemis、Lombok 等依赖，代码如例 5-15 所示。

【例 5-15】 添加 Web、ActiveMQ、Artemis、Lombok 依赖的代码示例。

```xml
<dependency>
        <groupId>org.springframework.boot</groupId>
        <artifactId>spring-boot-starter-web</artifactId>
</dependency>
<dependency>
        <groupId>org.springframework.boot</groupId>
        <artifactId>spring-boot-starter-activemq</artifactId>
</dependency>
<dependency>
        <groupId>org.springframework.boot</groupId>
        <artifactId>spring-boot-starter-artemis</artifactId>
</dependency>
<dependency>
        <groupId>org.projectlombok</groupId>
        <artifactId>lombok</artifactId>
        <version>1.18.8</version>
</dependency>
```

5.4.2 创建类 Block

在包 com.bookcode 中创建 entity 子包，并在包 com.bookcode.entity 中创建类 Block，代码如例 5-16 所示。

【例 5-16】 类 Block 的代码示例。

```
package com.bookcode.entity;
```

```
import com.bookcode.util.StringUtil;
import lombok.Data;
import java.util.Date;
@Data
public class Block {
private String hash;
private String previousHash;
private String data;
private long timestamp;
private int nonce;
public Block(String data, String previousHash) {
        super();
        this.previousHash = previousHash;
        this.data = data;
        this.timestamp = new Date().getTime();
        this.hash = calculateHash();
    }
    public String calculateHash() {
      String calculatedHash = StringUtil.applySHA256 ( previousHash + Long.toString
(timestamp) + data);
        return calculatedHash;
    }
    public String mineBlock() {
        //简化的模拟挖矿
        return "\n挖出一个新区块,其哈希值为:" + hash.toString();
    }
}
```

5.4.3 创建类 MQSendService

在包 com.bookcode 中创建 service 子包,并在包 com.bookcode.service 中创建类 MQSendService,代码如例 5-17 所示。

【例 5-17】 类 MQSendService 的代码示例。

```
package com.bookcode.service;
import org.springframework.beans.factory.annotation.Autowired;
import org.springframework.jms.annotation.JmsListener;
import org.springframework.jms.core.JmsMessagingTemplate;
import org.springframework.stereotype.Service;
import javax.jms.Destination;
//ActiveMQ 队列消息的生产者,即挖矿结果的报告者
@Service
public class MQSendService {
    @Autowired
    private JmsMessagingTemplate jmsMessagingTemplate;
    public void sendMessage(String msg, Destination destination){
        this.jmsMessagingTemplate.convertAndSend(destination, msg);
```

```
    }
    //双向队列
    @JmsListener(destination = "out.queue")
    public void consumerMessage(String text){
        System.out.println("从 out.queue 队列收到的回复报文为:" + text);
    }
}
```

5.4.4 创建类 MQReceiveService

在包 com.bookcode.service 中创建类 MQReceiveService,代码如例 5-18 所示。

【例 5-18】 类 MQReceiveService 的代码示例。

```
package com.bookcode.service;
import org.springframework.jms.annotation.JmsListener;
import org.springframework.jms.core.JmsMessagingTemplate;
import org.springframework.messaging.handler.annotation.SendTo;
import org.springframework.stereotype.Service;
//ActiveMQ 队列消息的生产者,即挖矿结果的接收者
@Service
public class MQReceiveService {
    private JmsMessagingTemplate jmsMessagingTemplate;
    //双向队列
    @JmsListener(destination = "BlockchainInfo.queue")
    @SendTo("out.queue")
    public String receiveBlockchainQueue(String text) {
        System.err.println("\n 收到的区块链消息如下: \n" + text);
        return "返回的消息如下: " + text;
    }
    //单向队列
    @JmsListener(destination = "MineInfo.queue")
    public void receiveMineInfoQueue(String text) {
        System.err.println("\n 挖矿的消息为: \n" + text);
    }
}
```

5.4.5 创建类 StringUtil

在包 com.bookcode 中创建 utils 子包,并在包 com.bookcode.utils 中创建类 StringUtil,代码如例 5-19 所示。

【例 5-19】 类 StringUtil 的代码示例。

```
package com.bookcode.utils;
import java.io.UnsupportedEncodingException;
```

```java
import java.security.MessageDigest;
import java.security.NoSuchAlgorithmException;
public class StringUtil{
    public static String applySHA256(String input) {
        StringBuffer hexString = null;
        try {
            MessageDigest messageDigest = MessageDigest.getInstance("SHA-256");
            byte[] hash = messageDigest.digest(input.getBytes("UTF-8"));
            hexString = new StringBuffer();
            String hex = null;
            for(int t = 0 ; t < hash.length ; t++) {
                hex = Integer.toHexString(0xff & hash[t]);
                if( hex.length() == 1 ) {
                    hexString.append('0');
                } else {
                    hexString.append(hex);
                }
            }
        } catch (NoSuchAlgorithmException e) {
            e.printStackTrace();
        } catch (UnsupportedEncodingException e) {
            e.printStackTrace();
        }
        return hexString.toString();
    }
}
```

5.4.6 创建类 SendInfoController

在包 com.bookcode 中创建 controller 子包，并在包 com.bookcode.controller 中创建类 SendInfoController，代码如例 5-20 所示。

【例 5-20】 类 SendInfoController 的代码示例。

```java
package com.bookcode.controller;
import com.bookcode.service.MQSendService;
import com.bookcode.entity.Block;
import org.apache.activemq.command.ActiveMQQueue;
import org.springframework.beans.factory.annotation.Autowired;
import org.springframework.web.bind.annotation.*;
import javax.jms.Destination;
import java.util.ArrayList;
@RestController
public class SendInfoController {
    @Autowired
    MQSendService mqSendService;
    ArrayList<Block> blockChains = new ArrayList<>();
    @GetMapping("/sendBCinfo")
```

```java
    public String sendBCMsg(){
        blockChains.add(new Block("第一个区块", "0"));
        blockChains.add(new Block("第二个区块", blockChains.get(blockChains.size()-1).calculateHash()));
        blockChains.add(new Block("第三个区块", blockChains.get(blockChains.size()-1).calculateHash()));
        blockChains.add(new Block("第四个区块", blockChains.get(blockChains.size()-1).calculateHash()));
        String strBlockchain = blockChains.toString();
        Destination destination = new ActiveMQQueue("BlockchainInfo.queue");
        mqSendService.sendMessage(strBlockchain, destination);
        return "区块链消息通过ActiveMQ发送成功.";
    }
    @GetMapping("/sendmineinfo")
    public String sendMineMsg(){
        String strMine = "";
        for(int i=0;i<blockChains.size();i++){
            strMine += blockChains.get(i).mineBlock();
        }
        Destination destination = new ActiveMQQueue("MineInfo.queue");
        mqSendService.sendMessage(strMine, destination);
        return "挖矿消息通过ActiveMQ发送成功.";
    }
}
```

5.4.7 修改配置文件 application.properties

修改配置文件 application.properties，修改后的代码如例 5-21 所示。

【例 5-21】 修改后的配置文件 application.properties 代码示例。

```
# true 值表示使用内置的 ActivMQ,false 值表示使用外置的 ActivMQ
spring.activemq.in-memory=true
# 是否在回滚消息之前停止消息传递.当启用此命令时,消息顺序不会被保留
spring.activemq.non-blocking-redelivery=false
# 假如使用外置的 ActivMQ,除了将 spring.activemq.in-memory 设置为 false 外还要进行下面的设置
# MQ 所在的服务器的地址
# spring.activemq.broker-url=tcp://127.0.0.1:61616
# spring.activemq.password=admin
# spring.activemq.user=admin
```

5.4.8 运行程序

运行程序后，先在浏览器中输入 localhost:8080/sendBCinfo,浏览器的输结果如图 5-11 所示,控制台的输出结果如图 5-12 所示。再在浏览器中输入 localhost:8080/sendmineinfo, 浏览器的输出结果如图 5-13 所示,控制台的输出结果如图 5-14 所示。请注意,由于本示例

中有两个队列(一个是单向队列,一个是双向队列),假如在浏览器中输出顺序相反,就会抛出异常。

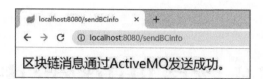

图 5-11　在浏览器中输入 localhost:8080/sendBCinfo 后浏览器的输出结果

```
收到的区块链消息如下:
[Block(hash=651e8a36c1ea4e2b8efa4ad25bee4a65f68ab080125160de3cccf5eac101075, previousHash=0, data=第一个区块,
从out.queue队列收到的回复报文为:返回的消息如下: [Block(hash=651e8a36c1ea4e2b8efa4ad25bee4a65f68ab080125160de3
```

图 5-12　在浏览器中输入 localhost:8080/sendBCinfo 后控制台的输出结果

图 5-13　在浏览器中输入 localhost:8080/sendmineinfo 后浏览器的输出结果

```
挖矿的消息为:
挖出一个新区块, 其哈希值为: 0b76bca61977b278fe6e41d75e81e4eedf14ca8a5c63f0a6a45ce4ccf7f5ba
挖出一个新区块, 其哈希值为: 336d1db82a9a36eb649d5a0bb50f73671a69411a14f2aafa87f9dcaa39480a8
挖出一个新区块, 其哈希值为: 918b49b3ed18ebaac651e2172eea434c9c1ca59ad8332b352d8b38b7fdf73cfe
挖出一个新区块, 其哈希值为: 5e4a209ec514adbc88a81f560a19d591b89bad3f24409866c4a5630c59b5a
```

图 5-14　在浏览器中输入 localhost:8080/sendmineinfo 后控制台的输出结果

习题 5

实验题

1. 实现 ActiveMQ 的应用。
2. 实现 RabbitMQ 的应用。
3. 实现 Spring Cloud Stream 和 RabbitMQ 的综合应用。

第6章 区块链应用中数据库开发

本章结合示例介绍如何在区块链应用中进行 MySQL、CouchDB 和 MongoDB 等数据库的开发。

6.1 MySQL 的应用

本节介绍 MySQL 在区块链中的应用开发。

视频讲解

6.1.1 创建项目并添加依赖

用 IDEA 创建完项目 hyperlfexample 之后,确保在文件 pom.xml 中<dependencies>和</dependencies>之间添加了 Web、bitcoinj、MySQL 驱动程序和 Spring Data JPA 相关依赖,代码如例 6-1 所示。

【例 6-1】 添加 Web、bitcoinj、MySQL 驱动程序和 Spring Data JPA 相关依赖的代码示例。

```
<dependency>
    <groupId>org.springframework.boot</groupId>
    <artifactId>spring-boot-starter-web</artifactId>
</dependency>
<dependency>
    <groupId>org.bitcoinj</groupId>
    <artifactId>bitcoinj-core</artifactId>
    <version>0.14.7</version>
</dependency>
<dependency>
    <groupId>mysql</groupId>
    <artifactId>mysql-connector-java</artifactId>
```

```
</dependency>
<dependency>
            <groupId>org.springframework.boot</groupId>
            <artifactId>spring-boot-starter-data-jpa</artifactId>
</dependency>
```

6.1.2 创建类 Pair

在包 com.bookcode 中创建 entity 子包,并在包 com.bookcode.entity 中创建类 Pair,代码如例 6-2 所示。

【例 6-2】 类 Pair 的代码示例。

```
package com.bookcode.entity;
import org.hibernate.annotations.CreationTimestamp;
import javax.persistence.*;
import java.math.BigDecimal;
import java.util.Date;
@Entity
@Table
public class Pair {
    @Id
    @Column(name = "public_address")
    private String publicAddress;
    @Column(name = "private_key", nullable = true)
    private String privateKey;
    @Column(name = "balance", nullable = true, precision = 20, scale = 8)
    private BigDecimal balance;
    @CreationTimestamp
    @Temporal(TemporalType.TIMESTAMP)
    @Column(name = "create_at")
    private Date createAt;
    public String getPublicAddress() {
        return publicAddress;
    }
    public void setPublicAddress(String publicAddress) {
        this.publicAddress = publicAddress;
    }
    public String getPrivateKey() {
        return privateKey;
    }
    public void setPrivateKey(String privateKey) {
        this.privateKey = privateKey;
    }
    public BigDecimal getBalance() {
        return balance;
    }
    public void setBalance(BigDecimal balance) {
        this.balance = balance;
```

```
        }
}
```

6.1.3 创建接口 PairService

在包 com.bookcode 中创建 service 子包,并在包 com.bookcode.service 中创建接口 PairService,代码如例 6-3 所示。

【例 6-3】 接口 PairService 的代码示例。

```
package com.bookcode.service;
import com.bookcode.entity.Pair;
import java.math.BigDecimal;
public interface PairService {
    public Pair createPair() throws Exception;
    public Pair savePair(Pair pair);
    public Pair getPairWithBalance(String publicAddress);
    public BigDecimal getBalance(String publicAddress) throws Exception;
}
```

6.1.4 创建类 PairServiceImpl

在包 com.bookcode.service 中创建 impl 子包,并在包 com.bookcode.service.impl 中创建类 PairServiceImpl,代码如例 6-4 所示。

【例 6-4】 类 PairServiceImpl 的代码示例。

```
package com.bookcode.service.impl;
import com.bookcode.entity.Pair;
import com.bookcode.repository.PairRepository;
import com.bookcode.service.PairService;
import org.springframework.beans.factory.annotation.Autowired;
import org.springframework.beans.factory.annotation.Qualifier;
import org.springframework.stereotype.Service;
import java.io.BufferedReader;
import java.io.InputStreamReader;
import java.math.BigDecimal;
import java.net.URL;
import java.util.Optional;
import org.bitcoinj.core.Address;
import org.bitcoinj.core.ECKey;
import org.bitcoinj.params.MainNetParams;
@Service
public class PairServiceImpl implements PairService {
    private PairRepository pairRepository;
```

```java
@Override
public Pair createPair() throws Exception {
    ECKey key = new ECKey();
    Address address = new Address(MainNetParams.get(), key.getPubKeyHash());
    Pair pair = new Pair();
    pair.setPublicAddress(address.toBase58());
    pair.setPrivateKey(key.getPrivateKeyAsWiF(MainNetParams.get()));
    try {
        pair.setBalance(getBalance(pair.getPublicAddress()));
    } catch (Exception e) {
        e.printStackTrace();
    }
    return savePair(pair);
}
@Override
public Pair savePair(Pair pair) {
    return pairRepository.save(pair);
}
@Override
public Pair getPairWithBalance(String publicAddress) {
    Optional<Pair> pairOpt = pairRepository.findById(publicAddress);
    Pair pair = null;
    if (pairOpt.isPresent()) {
        pair = pairOpt.get();
    } else {
        pair = new Pair();
        pair.setPublicAddress(publicAddress);
    }
    return pair;
}
@Override
public BigDecimal getBalance(String publicAddress) throws Exception {
    String url = getBalanceCheckUrl(publicAddress);
    String balance;
    balance = readUrl(url);
    return new BigDecimal(balance).divide(new BigDecimal("100000000"));
}
@Autowired
@Qualifier("pairRepository")
public void setPairRepository(PairRepository pairRepository) {
    this.pairRepository = pairRepository;
}
private String readUrl(String urlStr) throws Exception {
    String result = null;
    URL url = new URL(urlStr);
    BufferedReader in = new BufferedReader(new InputStreamReader(url.openStream()));
    String inputLine;
    while ((inputLine = in.readLine()) != null)
        result = inputLine;
    in.close();
```

```
            return result == null ? "0" : result;
    }
    private String getBalanceCheckUrl(String publicAddress) {
        String url = "https://blockchain.info/q/addressbalance/".concat(publicAddress);
        return url;
    }
}
```

6.1.5　创建接口 PairRepository

在包 com.bookcode 中创建 repository 子包,并在包 com.bookcode.repository 中创建接口 PairRepository,代码如例 6-5 所示。

【例 6-5】　PairRepository 的代码示例。

```
package com.bookcode.repository;
import com.bookcode.entity.Pair;
import org.springframework.data.jpa.repository.JpaRepository;
import org.springframework.stereotype.Repository;
@Repository("pairRepository")
public interface PairRepository extends JpaRepository<Pair, String> {
}
```

6.1.6　创建类 PairController

在包 com.bookcode 中创建 controller 子包,在包 com.bookcode.controller 中创建类 PairController,代码如例 6-6 所示。

【例 6-6】　类 PairController 的代码示例。

```
package com.bookcode.controller;
import com.bookcode.entity.Pair;
import com.bookcode.service.PairService;
import org.springframework.beans.factory.annotation.Autowired;
import org.springframework.web.bind.annotation.*;
import java.util.ArrayList;
import java.util.List;
@RestController
public class PairController {
    @Autowired
    private PairService pairService;
    @GetMapping(value = "/createPair")
    public Pair createPair() {
        try {
            return pairService.createPair();
```

```java
        } catch (Exception ex) {
            ex.printStackTrace();
        }
        return null;
    }
    @GetMapping(value = "/createPairList/{id}")
    public List<Pair> createPairList(@PathVariable int id) {
        List<Pair> list = new ArrayList<>();
        try {
            for (int i = 0; i < id; i++) {
                list.add(pairService.createPair());
            }
        } catch (Exception ex) {
            ex.printStackTrace();
        }
        return list;
    }
    @GetMapping(value = "/getBalance")
    public Pair getBalance(@RequestParam(value = "publicAddress", required = true,
defaultValue = "1MXahvQ35WwNGq1c7qEzXR3CK52vj665ms") String publicAddress) {
        try {
            return pairService.getPairWithBalance(publicAddress);
        } catch (Exception ex) {
            ex.printStackTrace();
        }
        return null;
    }
}
```

6.1.7 修改配置文件 application.properties

修改配置文件 application.properties，修改后的代码如例 6-7 所示。

【例 6-7】 修改后的配置文件 application.properties 代码示例。

```
spring.datasource.driver-class-name=com.mysql.cj.jdbc.Driver
spring.datasource.username=root
spring.datasource.password=sa
spring.jpa.hibernate.ddl-auto=update
spring.jpa.database-platform=org.hibernate.dialect.MySQLDialect
spring.datasource.url=jdbc:mysql://localhost:3306/test?serverTimezone=GMT%2B8\
&useUnicode=true&characterEncoding=UTF-8&useSSL=false
```

6.1.8 运行程序

运行程序后，在浏览器中输入 localhost:8080/createPair，结果如图 6-1 所示。在浏览器中输入 localhost:8080/createPairList/3,同时创建 3 条记录,结果如图 6-2 所示。在浏览器中输入 localhost:8080/getBalance,结果如图 6-3 所示。

图 6-1 在浏览器中输入 localhost:8080/createPair 的结果

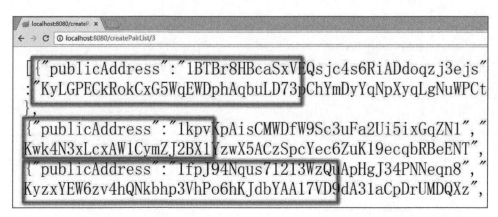

图 6-2 在浏览器中输入 localhost:8080/createPairList/3 的结果

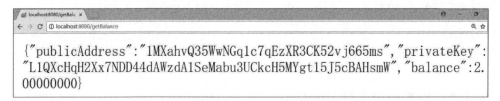

图 6-3 在浏览器中输入 localhost:8080/getBalance 的结果

6.2 CouchDB 的应用

视频讲解

由于 Fabric 使用了 CouchDB,故本节介绍 CouchDB 的应用。

6.2.1 创建项目并添加依赖

用 IDEA 创建完项目 cdb 之后,确保在文件 pom.xml 中<dependencies>和</dependencies>之间添加了 Ektorp、Web 等依赖,代码如例 6-8 所示。

【例 6-8】 添加 Ektorp、Web 等依赖的代码示例。

```
<dependency>
    <groupId>org.ektorp</groupId>
```

```xml
            <artifactId>org.ektorp</artifactId>
            <version>1.5.0</version>
</dependency>
<dependency>
            <groupId>org.ektorp</groupId>
            <artifactId>org.ektorp.spring</artifactId>
            <version>1.5.0</version>
</dependency>
<dependency>
            <groupId>io.swagger</groupId>
            <artifactId>swagger-annotations</artifactId>
            <version>1.5.20</version>
</dependency>
<dependency>
            <groupId>io.springfox</groupId>
            <artifactId>springfox-core</artifactId>
            <version>2.9.2</version>
</dependency>
<dependency>
            <groupId>io.springfox</groupId>
            <artifactId>springfox-swagger2</artifactId>
            <version>2.9.2</version>
</dependency>
<dependency>
            <groupId>org.springframework.boot</groupId>
            <artifactId>spring-boot-starter-web</artifactId>
</dependency>
```

6.2.2 创建类 CouchDBConfiguration

在包 com.bookcode 中创建 config 子包,并在包 com.bookcode.config 中创建类 CouchDBConfiguration,代码如例 6-9 所示。

【例 6-9】 类 CouchDBConfiguration 的代码示例。

```java
package com.bookcode.config;
import org.ektorp.CouchDbInstance;
import org.ektorp.DbAccessException;
import org.ektorp.http.HttpClient;
import org.ektorp.impl.StdCouchDbConnector;
import org.ektorp.impl.StdCouchDbInstance;
import org.ektorp.spring.HttpClientFactoryBean;
import org.slf4j.Logger;
import org.springframework.beans.factory.annotation.Value;
import org.springframework.context.annotation.Bean;
import org.springframework.context.annotation.Configuration;
import java.util.Properties;
import static org.slf4j.LoggerFactory.getLogger;
```

```java
@Configuration
public class CouchDBConfiguration {
    private static final Logger log = getLogger(CouchDBConfiguration.class);
    @Value("${mld.couchdb.host}")
    private String host;
    @Value("${mld.couchdb.port}")
    private String port;
    @Value("${mld.couchdb.database}")
    private String database;
    @Bean
    public StdCouchDbConnector connector() throws Exception {
log.info("Attempting to establish CouchDB Connector on {}:{}/{}",new Object[]{host,port,database});
        CouchDbInstance dbInstance = new StdCouchDbInstance(getHttpClient());
        StdCouchDbConnector connector = new StdCouchDbConnector(database, dbInstance);
        try {
            connector.createDatabaseIfNotExists();
        }catch (DbAccessException e) {
            log.warn("Cannot connect to the database (is the database up and available?)", e);
        }
        return connector;
    }
    private HttpClient getHttpClient() throws Exception {
        Properties properties = new Properties();
        properties.setProperty("autoUpdateViewOnChange", "true");
        HttpClientFactoryBean factory = new HttpClientFactoryBean();
        factory.setUrl(host + ":" + port);
        factory.setHost(host);
        factory.setPort(Integer.valueOf(port));
        factory.setProperties(properties);
        factory.afterPropertiesSet();
        return factory.getObject();
    }
}
```

6.2.3 创建类 Note

在包 com.bookcode 中创建 service 子包,并在包 com.bookcode.service 中创建类 Note,代码如例 6-10 所示。

【例 6-10】 类 Note 的代码示例。

```java
package com.bookcode.service;
import com.fasterxml.jackson.annotation.JsonCreator;
import com.fasterxml.jackson.annotation.JsonProperty;
import org.ektorp.support.CouchDbDocument;
public class Note extends CouchDbDocument {
    public String noteId;
```

```
    public String details;
    @JsonCreator
    public Note(@JsonProperty("noteId") String noteId, @JsonProperty("details") String details) {
        this.noteId = noteId;
        this.details = details;
        this.setId(noteId);
    }
    public String getNoteId() {
        return noteId;
    }
    public String getDetails() {
        return details;
    }
}
```

6.2.4 创建类 NotFoundException

在包 com.bookcode.service 中创建类 NotFoundException,代码如例 6-11 所示。

【例 6-11】 类 NotFoundException 的代码示例。

```
package com.bookcode.service;
public class NotFoundException extends RuntimeException {
    public NotFoundException(String message) {
        super(message);
    }
}
```

6.2.5 创建类 NotePersistenceHandler

在包 com.bookcode.service 中创建类 NotePersistenceHandler,代码如例 6-12 所示。

【例 6-12】 类 NotePersistenceHandler 的代码示例。

```
package com.bookcode.service;
import org.ektorp.CouchDbConnector;
import org.ektorp.support.CouchDbRepositorySupport;
import org.springframework.beans.factory.annotation.Autowired;
import org.springframework.stereotype.Component;
@Component
public class NotePersistenceHandler extends CouchDbRepositorySupport<Note> {
    @Autowired
    public NotePersistenceHandler(CouchDbConnector couchDbConnector) {
        super(Note.class, couchDbConnector);
        initStandardDesignDocument();
    }
}
```

6.2.6 创建类 NoteDTO

在包 com.bookcode 中创建 api 子包,并在 com.bookcode.api 包中创建 dto 子包,在包 com.bookcode.api.dto 中创建类 NoteDTO,代码如例 6-13 所示。

【例 6-13】 类 NoteDTO 的代码示例。

```java
package com.bookcode.api.dto;
import com.fasterxml.jackson.annotation.JsonCreator;
import com.fasterxml.jackson.annotation.JsonProperty;
import io.swagger.annotations.ApiModel;
@ApiModel(value = "NoteDTO")
public class NoteDTO {
    public String id;
    public String details;
    @JsonCreator
    public NoteDTO(@JsonProperty("id") String id, @JsonProperty("details") String details) {
        this.id = id;
        this.details = details;
    }
    public String getId() {
        return id;
    }
    public String getDetails() {
        return details;
    }
}
```

6.2.7 创建类 NoteService

在包 com.bookcode.service 中创建类 NoteService,代码如例 6-14 所示。

【例 6-14】 类 NoteService 的代码示例。

```java
package com.bookcode.service;
import com.bookcode.api.dto.NoteDTO;
import org.ektorp.DocumentNotFoundException;
import org.slf4j.Logger;
import org.springframework.beans.factory.annotation.Autowired;
import org.springframework.stereotype.Component;
import java.util.UUID;
import static org.slf4j.LoggerFactory.getLogger;
@Component
public class NoteService {
    private static final Logger log = getLogger(NoteService.class);
    @Autowired
```

```
    NotePersistenceHandler notePersistenceHandler;
    public NoteDTO createNote(String details){
        Note note = new Note(UUID.randomUUID().toString(), details);
        notePersistenceHandler.update(note);
        log.info("Note with id " + note.getId() + " created/updated.");
        return convertNoteToDTO(note);
    }
    public NoteDTO getNote(String id){
        Note note = null;
        try {
            note = notePersistenceHandler.get(id);
            }catch (DocumentNotFoundException dnfe){
            log.warn("Cannot find note with id : " + id);
            throw new NotFoundException("Cannot find note with id : " + id);
        }
         log.info("Found note with id: " + note.getId() + " with details: " + note.getDetails());
        return convertNoteToDTO(note);
    }
    private static NoteDTO convertNoteToDTO(Note note){
        return new NoteDTO(note.getId(), note.getDetails());
    }
}
```

6.2.8　创建类 NotesController

在包 com.bookcode.api 中创建类 NotesController，代码如例 6-15 所示。

【例 6-15】 类 NotesController 的代码示例。

```
package com.bookcode.api;
import com.bookcode.api.dto.NoteDTO;
import com.bookcode.service.NotFoundException;
import com.bookcode.service.NoteService;
import io.swagger.annotations.ApiOperation;
import org.slf4j.Logger;
import org.springframework.beans.factory.annotation.Autowired;
import org.springframework.http.ResponseEntity;
import org.springframework.web.bind.annotation.*;
import static org.slf4j.LoggerFactory.getLogger;
import static org.springframework.http.HttpStatus.INTERNAL_SERVER_ERROR;
import static org.springframework.http.HttpStatus.NOT_FOUND;
import static org.springframework.http.MediaType.APPLICATION_JSON_VALUE;
import static org.springframework.web.bind.annotation.RequestMethod.GET;
import static org.springframework.web.bind.annotation.RequestMethod.POST;
@RestController
public class NotesController {
    @Autowired
```

```
    private NoteService noteService;
    private static final Logger log = getLogger(NotesController.class);
    @ApiOperation(value = "Attempts to find a note by id", response = NoteDTO.class)
    @RequestMapping(value = "/", method = GET, produces = APPLICATION_JSON_VALUE)
    public ResponseEntity<NoteDTO> findNote(@RequestParam("id") String id) {
        return ResponseEntity.ok(noteService.getNote(id));
    }
    @ApiOperation(value = "Creates a new note", response = NoteDTO.class)
    @RequestMapping(value = "/", method = POST, produces = APPLICATION_JSON_VALUE)
    public ResponseEntity<NoteDTO> saveNote(@RequestBody() String details) {
        return ResponseEntity.ok(noteService.createNote(details));
    }
    @ExceptionHandler(Exception.class)
    @ResponseStatus(INTERNAL_SERVER_ERROR)
    public void handleException(Exception e) {
        log.warn("An unexpected error occurred", e);
    }
    @ExceptionHandler(NotFoundException.class)
    @ResponseStatus(NOT_FOUND)
    public void handleNotFoundException(Exception e) {
        log.warn("Could not find requested NoteDTO", e);
    }
}
```

6.2.9 修改配置文件 application.properties

修改配置文件 application.properties,修改后的代码如例 6-16 所示。

【例 6-16】 修改后的配置文件 application.properties 代码示例。

```
mld.couchdb.host = http://127.0.0.1
mld.couchdb.port = 5984
mld.couchdb.database = notesdb
```

6.2.10 修改入口类

修改入口类,修改后的入口类代码如例 6-17 所示。

【例 6-17】 修改后的入口类代码示例。

```
package com.bookcode;
import org.springframework.boot.SpringApplication;
import org.springframework.boot.autoconfigure.SpringBootApplication;
import org.slf4j.Logger;
import org.springframework.context.annotation.Bean;
import springfox.documentation.builders.PathSelectors;
import springfox.documentation.builders.RequestHandlerSelectors;
```

```
import springfox.documentation.spring.web.plugins.Docket;
import springfox.documentation.swagger2.annotations.EnableSwagger2;
import static org.slf4j.LoggerFactory.getLogger;
import static springfox.documentation.spi.DocumentationType.SWAGGER_2;
@SpringBootApplication
@EnableSwagger2
public class CdbApplication {
    private static final Logger log = getLogger(CdbApplication.class);
    @Bean
    public Docket api() {
        return new Docket(SWAGGER_2)
                .select()
                .apis(RequestHandlerSelectors.basePackage("com.bookcode.api"))
                .paths(PathSelectors.any())
                .build();
    }
    public static void main(String[] args) {
        SpringApplication.run(CdbApplication.class, args);
    }
}
```

6.2.11 运行程序

参考附录 H 安装 CouchDB 数据库，在 Windows 命令行程序中执行如例 6-18 所示的命令启动 CouchDB 数据库。

【例 6-18】 启动 CouchDB 数据库的命令示例。

```
couchdb
```

打开 Postman 工具，在 URL 处输入 http://127.0.0.1:8080，选择 POST 方法，单击 Send 按钮，显示状态 Status 为 OK，如图 6-4 所示。这样就会在数据库对应的表中增加一条记录。重复执行三次相同操作就可以在数据库对应的表中增加三条记录，结果如图 6-5 所示。在浏览器中输入 localhost:8080/? id=bc7aac8f-2bd4-4e5c-8376-e17036fcdbe3，查询数据库表得到的记录结果如图 6-6 所示。请注意，在执行此种操作时查询参数(id 值)需要调整。

图 6-4 在 Postman 中增加一条记录

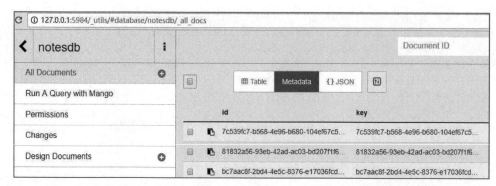

图 6-5　在 Postman 中执行三次 POST 操作后对应数据库增加了三条记录的结果

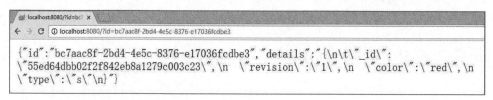

图 6-6　在浏览器中输入 localhost:8080/?id=bc7aac8f-2bd4-4e5c-8376-e17036fcdbe3 后的结果

6.3　MongoDB 的应用

视频讲解

　　在当前流行的 NoSQL 数据库中，MongoDB 数据库是用得比较多的数据库，MongoDB 数据库是基于文档的存储型数据库。MongoDB 数据库使用面向对象的思想，每条数据记录都是文档的对象。由于 Spring Boot 对 MongoDB 数据库有专门的支持，Spring Boot 在访问 MongoDB 数据库时不需要进行配置。使用 MongoDB 数据库前需要先安装 MongoDB 数据库，MongoDB 数据库有 Windows、Linux 等系统的安装包；安装过程比较简单，本书不作介绍。

6.3.1　创建项目并添加依赖

　　用 IDEA 创建完项目 exmongodb 之后，确保在 pom.xml 文件中＜dependencies＞和＜/dependencies＞之间添加了 MongoDB、Web 等依赖，代码如例 6-19 所示。

【例 6-19】　添加依赖的代码示例。

```
<dependency>
    <groupId>org.springframework.boot</groupId>
    <artifactId>spring-boot-starter-data-mongodb</artifactId>
</dependency>
<dependency>
    <groupId>org.springframework.boot</groupId>
    <artifactId>spring-boot-starter-web</artifactId>
</dependency>
```

```xml
<dependency>
            <groupId>org.bouncycastle</groupId>
            <artifactId>bcprov-jdk15on</artifactId>
            <version>1.60</version>
</dependency>
<dependency>
            <groupId>org.projectlombok</groupId>
            <artifactId>lombok</artifactId>
</dependency>
```

6.3.2 创建类 Block

在包 com.bookcode 中创建 entity 子包,并在包 com.bookcode.entity 中创建类 Block, 代码如例 6-20 所示。

【例 6-20】 类 Block 的代码示例。

```java
package com.bookcode.entity;
import lombok.Data;
import org.springframework.data.annotation.Id;
@Data
public class Block {
    @Id
    private String id;
    private String blockData;
    private String strSHA;
    public Block(String id, String blockData, String strSHA) {
        this.id = id;
        this.blockData = blockData;
        this.strSHA = strSHA;
    }
}
```

6.3.3 创建接口 BlockRepository

在包 com.bookcode 中创建 dao 子包,并在包 com.bookcode.dao 中创建接口 BlockRepository,代码如例 6-21 所示。接口 BlockRepository 的父接口 MongoRepository 封装了对数据库 MongoDB 的访问方法。

【例 6-21】 接口 BlockRepository 的代码示例。

```java
package com.bookcode.dao;
import com.bookcode.entity.Block;
import org.springframework.data.mongodb.repository.MongoRepository;
public interface BlockRepository extends MongoRepository<Block, String> {
}
```

6.3.4 创建类 SHA256

在包 com.bookcode 中创建 util 子包，并在包 com.bookcode.util 中创建辅助类 SHA256，代码如例 6-22 所示。

【例 6-22】 类 SHA256 的代码示例。

```
package com.bookcode.util;
import org.bouncycastle.util.encoders.Hex;
import java.nio.charset.StandardCharsets;
import java.security.MessageDigest;
import java.security.NoSuchAlgorithmException;
public class SHA256 {
    public static String setSha256hex(String originalString) {
        MessageDigest digest = null;
        try {
            digest = MessageDigest.getInstance("SHA");
        } catch (NoSuchAlgorithmException e) {
            e.printStackTrace();
        }
        byte[] hash = digest.digest(originalString.getBytes(StandardCharsets.UTF_8));
        return new String(Hex.encode(hash));
    }
}
```

6.3.5 创建类 PrimaryController

在包 com.bookcode 中创建 controller 子包，并在包 com.bookcode.controller 中创建类 PrimaryController，代码如例 6-23 所示。

【例 6-23】 类 PrimaryController 的代码示例。

```
package com.bookcode.controller;
import com.bookcode.entity.Block;
import com.bookcode.util.SHA256;
import org.springframework.beans.factory.annotation.Autowired;
import org.springframework.web.bind.annotation.*;
import java.util.List;
@RestController
public class PrimaryController {
    @Autowired
    BlockRepository blockRepository;
    @GetMapping("/lists")
    public List<Block> search(Block block){
        List<Block> blockList = blockRepository.findAll();
        return blockList;
    }
```

```
    @RequestMapping("/save/{id}/{blockData}")
    public Block save(@PathVariable String id,@PathVariable String blockData ) {
        String blockSHA = SHA256.setSha256hex(blockData);
        Block block = new Block(id,blockData,blockSHA);
        blockRepository.save(block);
        return block;
    }
}
```

6.3.6 修改配置文件 application. properties

修改配置文件 application. properties,代码如例 6-24 所示。

【例 6-24】 修改后的配置文件 application. properties 代码示例。

```
spring.data.mongodb.database = chain
spring.data.mongodb.host = localhost
spring.data.mongodb.port = 27017
```

6.3.7 运行程序

在 Windows 命令处理程序中输入命令,如例 6-25 所示,启动 MongoDB 数据库。代码中路径和 MongoDB 数据库的设置有关。

【例 6-25】 启动 MongoDB 数据库的命令示例。

```
mongodb -- dbpath "D:\mongodb\data\db"
```

启动 MongoDB 数据库后运行程序,再在浏览器中输入 localhost: 8080/save/1/zsblock,结果如图 6-7 所示。在浏览器中输入 localhost:8080/lists,结果如图 6-8 所示。

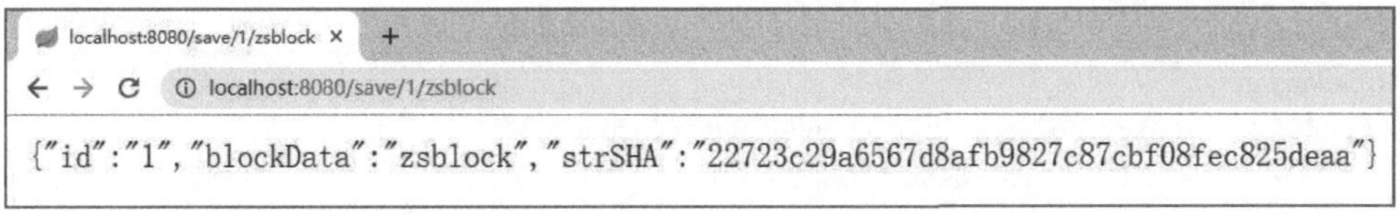
localhost:8080/save/1/zsblock
localhost:8080/save/1/zsblock
{"id":"1","blockData":"zsblock","strSHA":"22723c29a6567d8afb9827c87cbf08fec825deaa"}

图 6-7 在浏览器中输入 localhost:8080/save/1/zsblock 后的结果

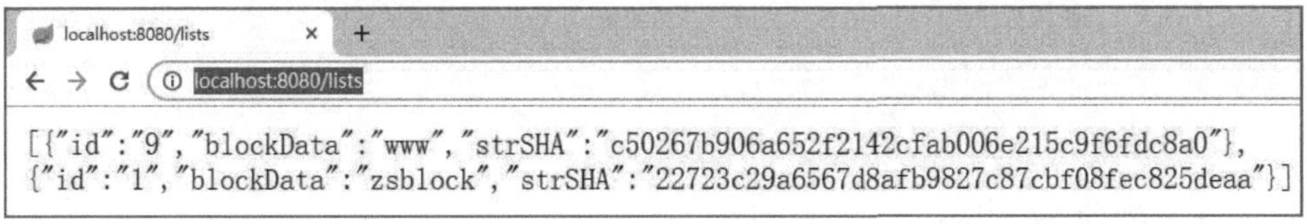
localhost:8080/lists
localhost:8080/lists
[{"id":"9","blockData":"www","strSHA":"c50267b906a652f2142cfab006e215c9f6fdc8a0"},
{"id":"1","blockData":"zsblock","strSHA":"22723c29a6567d8afb9827c87cbf08fec825deaa"}]

图 6-8 在浏览器中输入 localhost:8080/lists 后的结果

6.4 用以太坊区块链进行数据审核的示例

视频讲解

本节介绍利用 MongoDB 数据库和以太坊区块链 Ethereum 的 Web3j 进行数据审核的预订申请。

6.4.1 创建项目并添加依赖

用 IDEA 创建完项目 honesthomes 之后,确保在 pom.xml 文件中< dependencies >和</dependencies >之间添加了 MongoDB、Web 等依赖,代码如例 6-26 所示。

【例 6-26】 添加 MongoDB、Web 等依赖的代码示例。

```xml
<dependency>
    <groupId>org.springframework.boot</groupId>
    <artifactId>spring-boot-starter-data-mongodb</artifactId>
</dependency>
<dependency>
    <groupId>org.springframework.boot</groupId>
    <artifactId>spring-boot-starter-web</artifactId>
</dependency>
<dependency>
    <groupId>org.web3j</groupId>
    <artifactId>web3j-spring-boot-starter</artifactId>
</dependency>
<dependency>
    <groupId>org.web3j</groupId>
    <artifactId>core</artifactId>
    <version>2.3.0</version>
</dependency>
<dependency>
    <groupId>org.aspectj</groupId>
    <artifactId>aspectjweaver</artifactId>
</dependency>
<dependency>
    <groupId>com.google.code.gson</groupId>
    <artifactId>gson</artifactId>
</dependency>
<dependency>
    <groupId>org.projectlombok</groupId>
    <artifactId>lombok</artifactId>
</dependency>
```

6.4.2 创建类 User

在包 com.bookcode 中创建 entity 子包,并在包 com.bookcode.entity 中创建类 User,

代码如例 6-27 所示。

【例 6-27】 类 User 的代码示例。

```java
package com.bookcode.entity;
import org.springframework.data.annotation.Id;
import org.springframework.data.mongodb.core.mapping.Document;
import java.util.LinkedList;
import java.util.List;
@Document(collection = "customer")
public class User {
    @Id
    private String id;
    private String firstName;
    private String lastName;
    private List<String> ownedPropertyIds = new LinkedList<>();
    private List<String> visitedPropertyIds = new LinkedList<>();
    public String getId() {
        return id;
    }
    public void setId(String id) {
        this.id = id;
    }
    public String getFirstName() {
        return firstName;
    }
    public void setFirstName(String firstName) {
        this.firstName = firstName;
    }
    public String getLastName() {
        return lastName;
    }
    public void setLastName(String lastName) {
        this.lastName = lastName;
    }
    public List<String> getOwnedPropertyIds() {
        return ownedPropertyIds;
    }
    public void setOwnedPropertyIds(List<String> ownedPropertyIds) {
        this.ownedPropertyIds = ownedPropertyIds;
    }
    public boolean addOwnedPropertyId(String propertyId) {
        return ownedPropertyIds.add(propertyId);
    }
    public List<String> getVisitedPropertyIds() {
        return visitedPropertyIds;
    }
    public void setVisitedPropertyIds(List<String> visitedPropertyIds) {
        this.visitedPropertyIds = visitedPropertyIds;
    }
```

```java
    public boolean addVisitedPropertyId(String propertyId) {
        return visitedPropertyIds.add(propertyId);
    }
    @Override
    public String toString() {
        return "User{" +
                "id='" + id + '\'' +
                ", firstName='" + firstName + '\'' +
                ", lastName='" + lastName + '\'' +
                ", ownedPropertyIds=" + ownedPropertyIds +
                ", visitedPropertyIds=" + visitedPropertyIds +
                '}';
    }
}
```

6.4.3 创建类 Property

在包 com.bookcode.entity 中创建类 Property，代码如例 6-28 所示。

【例 6-28】 类 Property 的代码示例。

```java
package com.bookcode.entity;
import org.springframework.data.annotation.Id;
import org.springframework.data.mongodb.core.mapping.Document;
import java.util.LinkedList;
import java.util.List;
@Document(collection = "property")
public class Property {
    @Id
    private String id;
    private String landlordId;
    private String description;
    private List<String> imageUrls = new LinkedList<>();
    public String getId() {
        return id;
    }
    public void setId(String id) {
        this.id = id;
    }
    public String getLandlordId() {
        return landlordId;
    }
    public void setLandlordId(String landlordId) {
        this.landlordId = landlordId;
    }
    public String getDescription() {
        return description;
    }
```

```java
    public void setDescription(String description) {
        this.description = description;
    }
    public List<String> getImageUrls() {
        return imageUrls;
    }
    public void setImageUrls(List<String> imageUrls) {
        this.imageUrls = imageUrls;
    }
    public boolean addImageUrl(String imageUrl) {
        return imageUrls.add(imageUrl);
    }
    @Override
    public String toString() {
        return "Property{" +
                "id='" + id + '\'' +
                ", landlordId='" + landlordId + '\'' +
                ", description='" + description + '\'' +
                ", imageUrls=" + imageUrls +
                '}';
    }
}
```

6.4.4 创建类 Feedback

在包 com.bookcode.entity 中创建类 Feedback,代码如例 6-29 所示。

【例 6-29】 类 Feedback 的代码示例。

```java
package com.bookcode.entity;
import org.springframework.data.annotation.Id;
import org.springframework.data.mongodb.core.mapping.Document;
@Document(collection = "feedback")
public class Feedback {
    @Id
    private String id;
    private String authorId;
    private String entityId;
    private String description;
    private Integer score;
    public String getId() {
        return id;
    }
    public void setId(String id) {
        this.id = id;
    }
    public String getAuthorId() {
        return authorId;
    }
```

```
    }
    public void setAuthorId(String authorId) {
        this.authorId = authorId;
    }
    public String getEntityId() {
        return entityId;
    }
    public void setEntityId(String entityId) {
        this.entityId = entityId;
    }
    public String getDescription() {
        return description;
    }
    public void setDescription(String description) {
        this.description = description;
    }
    public Integer getScore() {
        return score;
    }
    public void setScore(Integer score) {
        this.score = score;
    }
    @Override
    public String toString() {
        return "Feedback{" +
                "id = '" + id + '\'' +
                ", authorId = '" + authorId + '\'' +
                ", entityId = '" + entityId + '\'' +
                ", description = '" + description + '\'' +
                ", score = " + score +
                '}';
    }
}
```

6.4.5　创建类 ContractDetails

在包 com.bookcode.entity 中创建类 ContractDetails，代码如例 6-30 所示。

【例 6-30】　类 ContractDetails 的代码示例。

```
package com.bookcode.entity;
import org.springframework.data.annotation.Id;
import org.springframework.data.mongodb.core.mapping.Document;
@Document(collection = "contract")
public class ContractDetails {
    @Id
    private String id;
    private String contractAddress;
```

```java
    public ContractDetails(String contractAddress) {
        this.contractAddress = contractAddress;
    }
    public String getId() {
        return id;
    }
    public void setId(String id) {
        this.id = id;
    }
    public String getContractAddress() {
        return contractAddress;
    }
    public void setContractAddress(String contractAddress) {
        this.contractAddress = contractAddress;
    }
    @Override
    public String toString() {
        return "ContractDetails{" +
                "id = '" + id + '\'' +
                ", contractAddress = '" + contractAddress + '\'' +
                '}';
    }
}
```

6.4.6 创建接口 Repository

在包 com.bookcode 中创建子包 dao，并在包 com.bookcode.dao 中创建接口 Repository，代码如例 6-31 所示。

【例 6-31】 接口 Repository 的代码示例。

```java
package com.bookcode.dao;
import org.springframework.data.mongodb.repository.MongoRepository;
import org.springframework.data.repository.NoRepositoryBean;
import java.io.Serializable;
@NoRepositoryBean
public interface Repository<T, ID extends Serializable> extends MongoRepository<T, ID> {
}
```

6.4.7 创建接口 UserRepository

在包 com.bookcode.dao 中创建接口 UserRepository，代码如例 6-32 所示。

【例 6-32】 接口 UserRepository 的代码示例。

```java
package com.bookcode.dao;
import com.bookcode.entity.User;
```

```
import java.util.List;
public interface UserRepository extends Repository<User, String> {
    List<User> findByLastName(String lastName);
}
```

6.4.8 创建接口 PropertyRepository

在包 com.bookcode.dao 中创建接口 PropertyRepository,代码如例 6-33 所示。

【例 6-33】 接口 PropertyRepository 的代码示例。

```
package com.bookcode.dao;
import com.bookcode.entity.Property;
import java.util.List;
public interface PropertyRepository extends Repository<Property, String> {
    List<Property> findByLandlordId(String landlordId);
}
```

6.4.9 创建接口 FeedbackRepository

在包 com.bookcode.dao 中创建接口 FeedbackRepository,代码如例 6-34 所示。

【例 6-34】 接口 FeedbackRepository 的代码示例。

```
package com.bookcode.dao;
import com.bookcode.entity.Feedback;
import java.util.List;
public interface FeedbackRepository extends Repository<Feedback, String> {
    List<Feedback> findByAuthorId(String authorId);
    List<Feedback> findByEntityId(String entityId);
}
```

6.4.10 创建接口 EthereumContractRepository

在包 com.bookcode.dao 中创建接口 EthereumContractRepository,代码如例 6-35 所示。

【例 6-35】 接口 EthereumContractRepository 的代码示例。

```
package com.bookcode.dao;
import com.bookcode.entity.ContractDetails;
public interface EthereumContractRepository extends Repository<ContractDetails, String> {
}
```

6.4.11 创建类 AbstractService

在包 com.bookcode 中创建 service 子包,并在包 com.bookcode.service 中创建类 AbstractService,代码如例 6-36 所示。

【例 6-36】 类 AbstractService 的代码示例。

```java
package com.bookcode.service;
import com.bookcode.dao.Repository;
import java.io.Serializable;
import java.util.List;
public abstract class AbstractService<T, ID extends Serializable> {
    protected Repository<T, ID> repository;
    public AbstractService(Repository<T, ID> repository) {
        this.repository = repository;
    }
    public List<T> findAll() {
        return repository.findAll();
    }
    public T findOne(ID id) {
        return repository.findById(id).get();
    }
    public T save(T entity) {
        return repository.save(entity);
    }
    public boolean exists(ID id) {
        return repository.existsById(id);
    }
    public boolean delete(ID id) {
        if (repository.existsById(id)) {
            repository.deleteById(id);
            return true;
        }
        return false;
    }
}
```

6.4.12 创建类 UserService

在包 com.bookcode.service 中创建类 UserService,代码如例 6-37 所示。

【例 6-37】 类 UserService 的代码示例。

```java
package com.bookcode.service;
import com.bookcode.dao.UserRepository;
import com.bookcode.entity.User;
import org.springframework.beans.factory.annotation.Autowired;
```

```java
import org.springframework.stereotype.Service;
import java.util.List;
@Service
public class UserService extends AbstractService<User, String> {
    @Autowired
    public UserService(UserRepository userRepository) {
        super(userRepository);
    }
    public List<User> findByLastName(String lastName) {
        return getRepositoryInstance().findByLastName(lastName);
    }
    private UserRepository getRepositoryInstance() {
        return (UserRepository) repository;
    }
}
```

6.4.13　创建类 PropertyService

在包 com.bookcode.service 中创建类 PropertyService，代码如例 6-38 所示。

【例 6-38】　类 PropertyService 的代码示例。

```java
package com.bookcode.service;
import com.bookcode.dao.PropertyRepository;
import com.bookcode.entity.Property;
import com.bookcode.entity.User;
import org.springframework.beans.factory.annotation.Autowired;
import org.springframework.stereotype.Service;
import java.util.List;
@SuppressWarnings("ALL")
@Service
public class PropertyService extends AbstractService<Property, String> {
    private UserService userService;
    @Autowired
    public PropertyService(PropertyRepository repository, UserService userService) {
        super(repository);
        this.userService = userService;
    }
    public List<Property> findByLandlordId(String landlordId) {
        return getRepositoryInstance().findByLandlordId(landlordId);
    }
    @Override
    public Property save(Property property) {
        if (property.getLandlordId() == null) {
            throw new RuntimeException("Cannot add property without owner. " +
                    "Landlord id must be provided.");
        }
        User owner = userService.findOne(property.getLandlordId());
```

```java
        if (owner == null) {
            throw new RuntimeException("Owner not found.");
        }
        Property saved = super.save(property);
        owner.addOwnedPropertyId(saved.getId());
        userService.save(owner);
        return saved;
    }
    private PropertyRepository getRepositoryInstance() {
        return (PropertyRepository) repository;
    }
}
```

6.4.14 创建类 FeedbackService

在包 com.bookcode.service 中创建类 FeedbackService，代码如例 6-39 所示。

【例 6-39】 类 FeedbackService 的代码示例。

```java
package com.bookcode.service;
import com.bookcode.dao.FeedbackRepository;
import com.bookcode.entity.Feedback;
import org.springframework.beans.factory.annotation.Autowired;
import org.springframework.stereotype.Service;
import javax.annotation.Resource;
import java.util.List;
@Service
public class FeedbackService extends AbstractService<Feedback, String> {
    @Resource
    EthereumService ethereumService;
    @Autowired
    public FeedbackService(FeedbackRepository repository, EthereumService ethereumService) {
        super(repository);
        this.ethereumService = ethereumService;
    }
    public List<Feedback> findByAuthorId(String authorId) {
        return getRepositoryInstance().findByAuthorId(authorId);
    }
    public List<Feedback> findByEntityId(String entityId) {
        return getRepositoryInstance().findByEntityId(entityId);
    }
    @Override
    public Feedback save(Feedback feedback) {
        feedback = super.save(feedback);
        ethereumService.storeFeedbackSha256Hex(feedback);
        return super.save(feedback);
    }
    private FeedbackRepository getRepositoryInstance() {
```

```
            return (FeedbackRepository) repository;
    }
}
```

6.4.15 创建类 EthereumService

在包 com.bookcode.service 中创建类 EthereumService，代码如例 6-40 所示。

【例 6-40】 类 EthereumService 的代码示例。

```
package com.bookcode.service;
import com.bookcode.dao.EthereumContractRepository;
import com.bookcode.entity.ContractDetails;
import com.bookcode.entity.Feedback;
import com.bookcode.utils.HashArray;
import com.bookcode.utils.Sha256Hex;
import com.google.gson.GsonBuilder;
import org.apache.commons.codec.digest.DigestUtils;
import org.slf4j.Logger;
import org.slf4j.LoggerFactory;
import org.springframework.beans.factory.annotation.Autowired;
import org.springframework.core.env.Environment;
import org.springframework.stereotype.Service;
import org.springframework.util.StringUtils;
import org.web3j.abi.datatypes.Utf8String;
import org.web3j.crypto.Credentials;
import org.web3j.protocol.Web3j;
import java.math.BigInteger;
import java.util.LinkedList;
import java.util.List;
import java.util.concurrent.ExecutionException;
import java.util.concurrent.Future;
import static org.web3j.tx.Contract.GAS_LIMIT;
import static org.web3j.tx.ManagedTransaction.GAS_PRICE;
@Service
public class EthereumService {
    private static final Logger LOGGER = LoggerFactory.getLogger(EthereumService.class);
    private HashArray contract;
    private EthereumContractRepository ethereumContractRepository;
    @Autowired
    public EthereumService(Environment environment, Credentials credentials, Web3j web3j,
EthereumContractRepository ethereumContractRepository) {
        this.ethereumContractRepository = ethereumContractRepository;
        initContract(environment, credentials, web3j, ethereumContractRepository);
    }
    public void storeFeedbackSha256Hex(Feedback feedback) {
        validate(feedback);
        String sha256Hex = calculateSha256Hex(feedback);
```

```java
        LOGGER.info("Storing sha256 hash: storeFeedbackSha256Hex(\"" + sha256Hex + "\")");
        Sha256Hex hash = new Sha256Hex(feedback.getId(), sha256Hex);
        contract.addHash(new Utf8String(new GsonBuilder().create().toJson(hash)));
    }
    public List<Sha256Hex> readFullFeedbackSha256HexList() {
        //模拟简化
        List<Sha256Hex> hexList = new LinkedList<>();
        Sha256Hex sha256Hex = new Sha256Hex("0001", "sha256001");
        hexList.add(sha256Hex);
        Sha256Hex sha256Hex1 = new Sha256Hex("0002", "sha256002");
        hexList.add(sha256Hex1);
        return hexList;
    }
    private String calculateSha256Hex(Feedback feedback) {
        String jsonString = new GsonBuilder().create().toJson(feedback);
        String sha256Hex = DigestUtils.sha256Hex(jsonString);
        LOGGER.info("Feedback: " + feedback.toString());
        LOGGER.info("Json: " + jsonString);
        LOGGER.info("Sha256Hex: " + sha256Hex);
        return sha256Hex;
    }
    private void validate(Feedback feedback) {
        if (StringUtils.isEmpty(feedback.getId())) {
            throw new RuntimeException("Feedback must be stored and have id assigned.");
        }
    }
}
private void initContract(Environment environment, Credentials credentials, Web3j web3j, EthereumContractRepository ethereumContractRepository) {
        String contractAddress = getDeployedContractAddress(environment, ethereumContractRepository);
        if (contractAddress == null) {
            //模拟简化
contract = HashArray.load("44da944df190a74a93b2bf769f53e7d3318de0a5", web3j, credentials, GAS_PRICE, GAS_LIMIT);
        } else {
contract = HashArray.load(contractAddress, web3j, credentials, GAS_PRICE, GAS_LIMIT);
        }
    }
    private String getDeployedContractAddress(Environment environment, EthereumContractRepository ethereumContractRepository) {
        String contractAddress = environment.getProperty("eth.geth.contract.address");
        if (contractAddress == null) {
            List<ContractDetails> all = ethereumContractRepository.findAll();
            if (all.size() == 1) {
                contractAddress = all.get(0).getContractAddress();
            }
            if (all.size() > 1) {
                throw new RuntimeException("Expected 1 contract record, but got " + all.size() + ". The contract should only be deployed only once.");
```

```
            }
        }
        return contractAddress;
    }
    private HashArray deployContract(Web3j web3j, Credentials credentials) {
        HashArray contract;
Future < HashArray > deploy = HashArray.deploy(web3j, credentials, GAS_PRICE, GAS_LIMIT, BigInteger.ZERO);
        try {
            contract = deploy.get();
            //模拟简化
            String strCD = "44da944df190a74a93b2bf769f53e7d3318de0a5";
            ethereumContractRepository.save(new ContractDetails(strCD));
        } catch (InterruptedException | ExecutionException e) {
            LOGGER.error("Error deploying contract. " + e.getMessage());
            throw new RuntimeException("Error deploying contract.", e);
        }
        return contract;
    }
}
```

6.4.16　创建类 ControllerLoggingAspect

在包 com.bookcode 中创建 aspect 子包，并在包 com.bookcode.aspect 中创建类 ControllerLoggingAspect，代码如例 6-41 所示。

【例 6-41】　类 ControllerLoggingAspect 的代码示例。

```
package com.bookcode.aspect;
import org.aspectj.lang.JoinPoint;
import org.aspectj.lang.annotation.Aspect;
import org.aspectj.lang.annotation.Before;
import org.slf4j.Logger;
import org.slf4j.LoggerFactory;
import org.springframework.stereotype.Component;
@Aspect
@Component
public class ControllerLoggingAspect {
    private static final Logger LOGGER = LoggerFactory.getLogger(ControllerLoggingAspect.class);
    @Before("execution( * com.bookcode.controller..*.*(..))")
    public void logBefore(JoinPoint joinPoint) {
        String argScv = "";
        Object[] signatureArgs = joinPoint.getArgs();
        if (signatureArgs != null && signatureArgs.length > 0) {
            StringBuilder args = new StringBuilder();
            for (Object signatureArg : signatureArgs) {
                args.append(signatureArg.toString()).append(",");
```

```
        }
        argScv = args.substring(0, args.length() - 1);
    }
    LOGGER.info("REST CALL >> " + joinPoint.getSignature().getDeclaringTypeName() +
        "." + joinPoint.getSignature().getName() + "(" + argScv + ")");
    }
}
```

6.4.17 创建类 Sha256Hex

在包 com.bookcode 中创建 utils 子包,并在包 com.bookcode.utils 中创建类 Sha256Hex,代码如例 6-42 所示。

【例 6-42】 类 Sha256Hex 的代码示例。

```
package com.bookcode.utils;
import lombok.AllArgsConstructor;
import lombok.Data;
@Data
@AllArgsConstructor
public class Sha256Hex {
    private String feedbackId;
    private String sha256Hex;
}
```

6.4.18 创建类 HashArray

在包 com.bookcode.utils 中创建类 HashArray,代码如例 6-43 所示。

【例 6-43】 类 HashArray 的代码示例。

```
package com.bookcode.utils;
import org.web3j.abi.TypeReference;
import org.web3j.abi.datatypes.Function;
import org.web3j.abi.datatypes.Type;
import org.web3j.abi.datatypes.Utf8String;
import org.web3j.abi.datatypes.generated.Uint256;
import org.web3j.crypto.Credentials;
import org.web3j.protocol.Web3j;
import org.web3j.protocol.core.methods.response.TransactionReceipt;
import org.web3j.tx.Contract;
import org.web3j.tx.TransactionManager;
import java.math.BigInteger;
import java.util.Arrays;
import java.util.Collections;
import java.util.concurrent.Future;
```

```java
public final class HashArray extends Contract {
    private static final String BINARY = "6060604052341561000f57600080fd5b5b6105478061001
f6000396000f300606060405263ffffffff7c0100000000000000000000000000000000000000000000
00000000060003504166314155b022811461005e5780636b2fafa9146100ec578063aeb276021461017a57806
3d53bdf05146101dd575b600080fd5b34156100695760008 0fd5b610074600435610202565b604051602080
82528190810183818151815260200191508051906020019080838360005b8381101561 0b15780820151818
401525b602001610098565b5050505090509081019060 1f1680156100de578082038051600183602003610
000a031916815260200191505b50925050506040518091039 0f35b3415610 0f757600080fd5b6100746004 3
56102be565b604051602080 8252819081018381815182620 0191508051906020019080838360 05b8381
10156100b15780820151818401525b602001610098565b5050505 09050908101906 01f1680156 100de57808
20380516001836020036101000a03 1916815260200191505b509250 50506040518091039 0f35b3415610185
57600080fd5b6101cb60046024813581010 9083013580602 0601f82018190048102016040 5190810160405
2818152929190602084018383808284375094965061038 495505050 5050505b6604051908 152602001604 0
5180910390f35b34156101e857600080fd5b6101cb610 3c6565b604051908152602002060 160405180910390f35
b60008054829081106102105 7fe5b906000526020600020900160005b9150 90508054600181600116156101
000203166002900480601f0160208091040260200160405190810160405280929190 818152602001828 0546
00181600116156101000203166002900480156102b65780601f1061028b57610 1008083540402835291602 0
01916102b6565b82019190600052602060 002090 05b81548152906 0010190602001808311610299578 290036
01f168201915b505050505 081565b6102c66103cd565b60008054839081106102d457fe5b906 00052602060
0020900160005b508054600181600116156101000203166002900480601f01602 080910402620 0160405 19
081016040528092919081815260200182805460018160011615610100020316600290 04801561 03 77578060
01f1061034c576101008083540402835291602 001 91610377565b820191906000 5260206 00020905b81548 15
2906001019060 0200180831161035a57829003601f168201915b505050 505090 505b919050565b6 0006 0016
008054806001018281610 39c91906103df565b9160005260206000 20900160005b50848051610 3bc92916 02
00190610409565b5003905 05b919050565b6000545909565b602060405190810160405260008152 90565b 81
548183558181151161043576000838152 60209 02610 4039181019 0830161048 8565b5b505050 565b82805
4600181600116156101000203166002 900490 60005260206000 2090601f0160209 00481019282601f106104
4a57805160ff191683800117855561047756 56b82800160001018 5558215610 47757918201 5b828111156104 7
7578251825591602001919060010 19061045c565b5b5060 61048492915061048e565b50905b6 b6b108 10 19059 05b
8082111561048457600061048 282826104a2826104d 3565b506 0010161048e565b509 05b90565b6103 ca9 19 05b80 8
2111561048457600081556001 016104 5760 006104a282 6104d3 565b506 0010 161048e565b5 0905b 4600 18160 0 16156101000203166002
90046000825580601f10610 4f95750 6105 1 7565b601f0160 2090040 9060005 26020600020 90810190 610 5179
1906104b2565b5b5050 600a165627 a7a7230582 09a23 9b0eda 09f5e82c6986e 48e4c883bc7ddf40e38837e86e
8cc556eec662a920029";
private HashArray(String contractAddress, Web3j web3j, Credentials credentials, BigInteger
gasPrice, BigInteger gasLimit) {
        super(BINARY, contractAddress, web3j, credentials, gasPrice, gasLimit);
    }
private HashArray ( String   contractAddress,   Web3j   web3j,   TransactionManager
transactionManager, BigInteger gasPrice, BigInteger gasLimit) {
        super(BINARY, contractAddress, web3j, transactionManager, gasPrice, gasLimit);
    }
public static Future < HashArray > deploy(Web3j web3j, Credentials credentials, BigInteger
gasPrice, BigInteger gasLimit, BigInteger initialWeiValue) {
        return deployAsync(HashArray.class, web3j, credentials, gasPrice, gasLimit, BINARY,
"", initialWeiValue);
    }
    public   static   Future  <  HashArray >  deploy ( Web3j   web3j,   TransactionManager
transactionManager, BigInteger gasPrice, BigInteger gasLimit, BigInteger initialWeiValue) {
```

```java
        return deployAsync(HashArray.class, web3j, transactionManager, gasPrice, gasLimit, BINARY,
"", initialWeiValue);
    }
    public static HashArray load ( String contractAddress, Web3j web3j, Credentials
credentials, BigInteger gasPrice, BigInteger gasLimit) {
        return new HashArray(contractAddress, web3j, credentials, gasPrice, gasLimit);
    }
    public static HashArray load(String contractAddress, Web3j web3j, TransactionManager
transactionManager, BigInteger gasPrice, BigInteger gasLimit) {
        return new HashArray (contractAddress, web3j, transactionManager, gasPrice,
gasLimit);
    }
    public Future<Utf8String> hashesArray(Uint256 param0) {
        Function function = new Function("hashesArray",
            Arrays.<Type>asList(param0),
            Arrays.<TypeReference<?>>asList(new TypeReference<Utf8String>() {
            }));
        return executeCallSingleValueReturnAsync(function);
    }
    public Future<Utf8String> getHash(Uint256 index) {
        Function function = new Function("getHash",
            Arrays.<Type>asList(index),
            Arrays.<TypeReference<?>>asList(new TypeReference<Utf8String>() {
            }));
        return executeCallSingleValueReturnAsync(function);
    }
    public Future<TransactionReceipt> addHash(Utf8String hash) {
Function function = new Function("addHash", Arrays.<Type>asList(hash), Collections.
<TypeReference<?>>emptyList());
        return executeTransactionAsync(function);
    }
    public Future<Uint256> getHashesCount() {
        Function function = new Function("getHashesCount",
            Arrays.<Type>asList(),
            Arrays.<TypeReference<?>>asList(new TypeReference<Uint256>() {
            }));
        return executeCallSingleValueReturnAsync(function);
    }
}
```

6.4.19 创建类 ApplicationConfig

在包 com.bookcode 中创建 config 子包，并在包 com.bookcode.config 中创建类 ApplicationConfig，代码如例 6-44 所示。

【例 6-44】 类 ApplicationConfig 的代码示例。

```java
package com.bookcode.config;
import org.slf4j.Logger;
import org.slf4j.LoggerFactory;
import org.springframework.context.annotation.Bean;
import org.springframework.context.annotation.Configuration;
import org.springframework.core.env.Environment;
import org.web3j.crypto.CipherException;
import org.web3j.crypto.Credentials;
import org.web3j.crypto.WalletUtils;
import org.web3j.protocol.Web3j;
import org.web3j.protocol.http.HttpService;
import javax.annotation.Resource;
import java.io.IOException;
@Configuration
public class ApplicationConfig {
    private static final Logger LOGGER = LoggerFactory.getLogger(ApplicationConfig.class);
    @Resource
    Environment environment;
    @Bean
    public HttpService httpService() {
        return new HttpService(environment.getProperty("eth.host.url"));
    }
    @Bean
    public Credentials credentials() {
        Credentials credentials = null;
        try {
            credentials = WalletUtils.loadCredentials(
                    environment.getProperty("eth.geth.account.password"),
                    environment.getProperty("eth.geth.account.file"));
        } catch (IOException | CipherException e) {
            LOGGER.error("Unable to load Ethereum credentials." + e.getMessage());
            throw new RuntimeException(e);
        }
        return credentials;
    }
    @Bean
    public Web3j web3j() {
        return Web3j.build(httpService());
    }
}
```

6.4.20 创建类 UserController

在包 com.bookcode 中创建 controller 子包，并在包 com.bookcode.controller 中创建类 UserController，代码如例 6-45 所示。

【例 6-45】 类 UserController 的代码示例。

```java
package com.bookcode.controller;
import com.bookcode.entity.User;
import com.bookcode.service.UserService;
import org.springframework.http.HttpStatus;
import org.springframework.http.ResponseEntity;
import org.springframework.stereotype.Controller;
import org.springframework.web.bind.annotation.PathVariable;
import org.springframework.web.bind.annotation.RequestBody;
import org.springframework.web.bind.annotation.RequestMapping;
import org.springframework.web.bind.annotation.RequestMethod;
import javax.annotation.Resource;
import java.util.List;
@Controller
@RequestMapping("/users")
public class UserController {
    @Resource
    UserService userService;
    @RequestMapping(value = "/", method = RequestMethod.GET, produces = "application/json")
    public ResponseEntity getAll() {
        return new ResponseEntity<List<User>>(userService.findAll(), HttpStatus.OK);
    }
@RequestMapping(value = "/id/{id}", method = RequestMethod.GET, produces = "application/json")
    public ResponseEntity findById(@PathVariable("id") String id) {
        return new ResponseEntity<User>(userService.findOne(id), HttpStatus.OK);
    }
@RequestMapping(value = "/lastName/{lastName}", method = RequestMethod.GET, produces = "application/json")
    public ResponseEntity findByLastName(@PathVariable("lastName") String lastName) {
return new ResponseEntity<List<User>>(userService.findByLastName(lastName), HttpStatus.OK);
    }
    @RequestMapping(value = "/", method = RequestMethod.POST, produces = "application/json")
    public ResponseEntity create(@RequestBody User user) {
        return new ResponseEntity<User>(userService.save(user), HttpStatus.CREATED);
    }
    @RequestMapping(value = "/", method = RequestMethod.PUT, produces = "application/json")
    public ResponseEntity update(@RequestBody User user) {
        return new ResponseEntity<User>(userService.save(user), HttpStatus.CREATED);
    }
@RequestMapping(value = "/{id}", method = RequestMethod.DELETE, produces = "application/json")
    public ResponseEntity delete(@PathVariable("id") String id) {
        HttpStatus status = HttpStatus.NOT_FOUND;
        if (userService.exists(id)) {
            userService.delete(id);
            status = HttpStatus.ACCEPTED;
```

```
        }
        return new ResponseEntity(status);
    }
}
```

6.4.21 创建类 PropertyController

在包 com.bookcode.controller 中创建类 PropertyController,代码如例 6-46 所示。

【例 6-46】 类 PropertyController 的代码示例。

```
package com.bookcode.controller;
import com.bookcode.entity.Property;
import com.bookcode.service.PropertyService;
import org.springframework.http.HttpStatus;
import org.springframework.http.ResponseEntity;
import org.springframework.stereotype.Controller;
import org.springframework.web.bind.annotation.PathVariable;
import org.springframework.web.bind.annotation.RequestBody;
import org.springframework.web.bind.annotation.RequestMapping;
import org.springframework.web.bind.annotation.RequestMethod;
import javax.annotation.Resource;
import java.util.List;
@Controller
@RequestMapping("/property")
public class PropertyController {
    @Resource
    PropertyService propertyService;
    @RequestMapping(value = "/", method = RequestMethod.GET, produces = "application/json")
    public ResponseEntity getAll() {
        return new ResponseEntity<List<Property>>(propertyService.findAll(), HttpStatus.OK);
    }
    @RequestMapping(value = "/id/{id}", method = RequestMethod.GET, produces = "application/json")
    public ResponseEntity findById(@PathVariable("id") String id) {
        return new ResponseEntity<Property>(propertyService.findOne(id), HttpStatus.OK);
    }
    @RequestMapping(value = "/landlordId/{landlordId}", method = RequestMethod.GET, produces = "application/json")
    public ResponseEntity findByLandlordId(@PathVariable("landlordId") String landlordId) {
        return new ResponseEntity<List<Property>>(propertyService.findByLandlordId(landlordId), HttpStatus.OK);
    }
    @RequestMapping(value = "/", method = RequestMethod.POST, produces = "application/json")
    public ResponseEntity create(@RequestBody Property property) {
```

```java
            return new ResponseEntity < Property >(propertyService.save(property), HttpStatus.CREATED);
    }
    @RequestMapping(value = "/", method = RequestMethod.PUT, produces = "application/json")
    public ResponseEntity update(@RequestBody Property property) {
            return new ResponseEntity < Property >(propertyService.save(property), HttpStatus.CREATED);
    }
@RequestMapping(value = "/{id}", method = RequestMethod.DELETE, produces = "application/json")
    public ResponseEntity delete(@PathVariable("id") String id) {
        HttpStatus status = HttpStatus.NOT_FOUND;
        if (propertyService.exists(id)) {
            propertyService.delete(id);
            status = HttpStatus.ACCEPTED;
        }
        return new ResponseEntity(status);
    }
}
```

6.4.22 创建类 FeedbackController

在包 com.bookcode.controller 中创建类 FeedbackController,代码如例 6-47 所示。

【例 6-47】 类 FeedbackController 的代码示例。

```java
package com.bookcode.controller;
import com.bookcode.entity.Feedback;
import com.bookcode.service.FeedbackService;
import org.springframework.http.HttpStatus;
import org.springframework.http.ResponseEntity;
import org.springframework.stereotype.Controller;
import org.springframework.web.bind.annotation.*;
import javax.annotation.Resource;
import java.util.List;
@Controller
@RequestMapping("/feedback")
public class FeedbackController {
    @Resource
    FeedbackService feedbackService;
    @GetMapping("/")
    public ResponseEntity getAll() {
            return new ResponseEntity < List < Feedback >>(feedbackService.findAll(), HttpStatus.OK);
    }
@RequestMapping(value = "/id/{id}", method = RequestMethod.GET, produces = "application/json")
```

```java
    public ResponseEntity findById(@PathVariable("id") String id) {
        return new ResponseEntity<Feedback>(feedbackService.findOne(id), HttpStatus.OK);
    }
    @RequestMapping(value = "/authorId/{authorId}", method = RequestMethod.GET, produces = "application/json")
    public ResponseEntity findByAuthorId(@PathVariable("authorId") String landlordId) {
        return new ResponseEntity<List<Feedback>>(feedbackService.findByAuthorId(landlordId), HttpStatus.OK);
    }
    @RequestMapping(value = "/entityId/{entityId}", method = RequestMethod.GET, produces = "application/json")
    public ResponseEntity findByEntityId(@PathVariable("authorId") String landlordId) {
        return new ResponseEntity<List<Feedback>>(feedbackService.findByEntityId(landlordId), HttpStatus.OK);
    }
    @RequestMapping(value = "/", method = RequestMethod.POST, produces = "application/json")
    public ResponseEntity create(@RequestBody Feedback property) {
        return new ResponseEntity<Feedback>(feedbackService.save(property), HttpStatus.CREATED);
    }
    @RequestMapping(value = "/", method = RequestMethod.PUT, produces = "application/json")
    public ResponseEntity update(@RequestBody Feedback property) {
        return new ResponseEntity<Feedback>(feedbackService.save(property), HttpStatus.CREATED);
    }
    @RequestMapping(value = "/{id}", method = RequestMethod.DELETE, produces = "application/json")
    public ResponseEntity delete(@PathVariable("id") String id) {
        HttpStatus status = HttpStatus.NOT_FOUND;
        if (feedbackService.exists(id)) {
            feedbackService.delete(id);
            status = HttpStatus.ACCEPTED;
        }
        return new ResponseEntity(status);
    }
}
```

6.4.23 创建类 EthereumController

在包 com.bookcode.controller 中创建类 EthereumController,代码如例 6-48 所示。

【例 6-48】 类 EthereumController 的代码示例。

```java
package com.bookcode.controller;
import com.bookcode.service.EthereumService;
import com.bookcode.utils.Sha256Hex;
import org.springframework.http.HttpStatus;
```

```
import org.springframework.http.ResponseEntity;
import org.springframework.stereotype.Controller;
import org.springframework.web.bind.annotation.GetMapping;
import org.springframework.web.bind.annotation.RequestMapping;
import org.springframework.web.bind.annotation.ResponseBody;
import javax.annotation.Resource;
import java.util.List;
@Controller
@RequestMapping("/sha256")
public class EthereumController {
    @Resource
    EthereumService ethereumService;
    @GetMapping("/")
    public ResponseEntity<List<Sha256Hex>> getAll() {
return new ResponseEntity<>(ethereumService.readFullFeedbackSha256HexList(), HttpStatus.OK);
    }
    @GetMapping("/home")
    @ResponseBody
    String home() {
        return "<p align=\"left\"><b>It's alive!</b></p>";
    }
}
```

6.4.24 修改配置文件 application.properties

修改配置文件 application.properties，修改后的代码如例 6-49 所示。

【例 6-49】 修改后的配置文件 application.properties 代码示例。

```
#Server
server.port=8080
server.address=127.0.0.1
#Mongodb
spring.data.mongodb.host=localhost
spring.data.mongodb.port=27017
spring.data.mongodb.database=honest-homes
#Logging
logging.level.org.springframework.web=INFO
logging.level.com.exadel=DEBUG
logging.file=${java.io.tmpdir}/application.log
#Ethereum
eth.host.url=http://localhost:8545/
#修改成用户自己的钱包和密码信息,请参考附录 A.2 来生成该信息
eth.geth.account.file=C:\Users\ws\AppData\Roaming\Electrum\UTC--2019-06-20T10-19-01.180Z--3b7335b84e8a13f049955b1340e10ecde7e3329b
eth.geth.account.password=bookcodews780125
```

6.4.25 运行程序

在 Windows 命令处理程序中输入命令,代码如例 6-25 所示,启动 MongoDB 数据库。双击 C:\Users\ws\AppData\Roaming\npm(请改成用户自己的安装目录)下的 ganache-cli.cmd 文件,启动 Ganache CLI 工具。

打开 Postman 工具,在 URL 处输入 http://127.0.0.1:8080/property/,选择 POST 方法,单击 Send 按钮,显示状态 Status 为 OK,如图 6-9 所示。在浏览器中输入 localhost:8080/property/,结果如图 6-10 所示。在浏览器中输入 localhost:8080/property/id/0003,结果如图 6-11 所示。在浏览器中输入 localhost:8080/sha256/,结果如图 6-12 所示。在浏览器中输入 localhost:8080/sha256/home,结果如图 6-13 所示。本处实现的操作(功能)较多,其他更多操作请根据源代码进行分析。

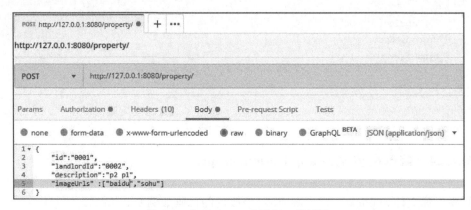

图 6-9 在 Postman 中增加一条记录

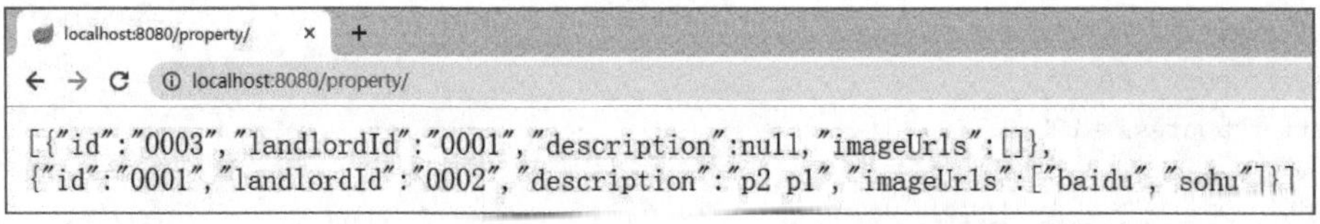

图 6-10 在浏览器中输入 localhost:8080/property/ 的结果

图 6-11 在浏览器中输入 localhost:8080/property/id/0003 的结果

图 6-12 在浏览器中输入 localhost:8080/sha256/ 的结果

图 6-13　在浏览器中输入 localhost：8080/sha256/home 的结果

习题 6

实验题

1. 实现 MySQL 的应用开发。
2. 实现 CouchDB 的应用开发。
3. 实现 MongoDB 的应用开发。

第7章

区块链应用中共识算法的实现

所谓共识,简单理解就是达成一致。在区块链系统中,每个节点必须要做的事情就是让自己的账本跟其他节点的账本保持一致。在区块链系统中,如何让每个节点通过一个规则将各自的数据保持一致是一个很核心的问题,这个问题的解决方案就是制定一套共识算法。在区块链系统中,已经存在PoW(Proof of Work,工作量证明)、PoS(Proof of Stake,权益证明)、DPoS(Delegate Proof of Stake,委托权益证明)、PBFT(Practical Byzantine Fault Tolerance,实用拜占庭容错算法)等不同的算法。本章介绍 PoW、PBFT、Raft 等共识算法的实现。

7.1 PoW 算法的实现

工作量证明简单理解就是一份证明,用来确认区块链中的节点做过一定量的工作。工作量证明要求发起者进行一定量的运算(工作),也就意味着需要消耗计算机一定的时间。

7.1.1 创建项目并添加依赖

用 IDEA 创建完项目 regswaggerexample 之后,确保在文件 pom.xml 中< dependencies >和</dependencies >之间添加了 Web、Actuator 等依赖,代码如例 7-1 所示。

【例 7-1】 添加 Web、Actuator 等依赖的代码示例。

```
< dependency >
        < groupId > org.springframework.boot </groupId >
        < artifactId > spring - boot - starter - actuator </artifactId >
</dependency >
< dependency >
        < groupId > org.springframework.boot </groupId >
```

```xml
            <artifactId>spring-boot-starter-web</artifactId>
</dependency>
<dependency>
            <groupId>org.projectlombok</groupId>
            <artifactId>lombok</artifactId>
            <version>1.18.4</version>
            <scope>provided</scope>
</dependency>
<dependency>
            <groupId>io.springfox</groupId>
            <artifactId>springfox-swagger-ui</artifactId>
            <version>2.9.2</version>
</dependency>
<dependency>
            <groupId>io.springfox</groupId>
            <artifactId>springfox-swagger2</artifactId>
            <version>2.9.2</version>
</dependency>
<dependency>
            <groupId>com.alibaba</groupId>
            <artifactId>fastjson</artifactId>
            <version>1.2.50</version>
</dependency>
<dependency>
            <groupId>org.apache.commons</groupId>
            <artifactId>commons-collections4</artifactId>
            <version>4.1</version>
</dependency>
<dependency>
            <groupId>commons-codec</groupId>
            <artifactId>commons-codec</artifactId>
            <version>1.11</version>
</dependency>
<dependency>
            <groupId>org.springframework.boot</groupId>
            <artifactId>spring-boot-starter-web</artifactId>
</dependency>
```

7.1.2 创建类 BeanInjector

在包 com.bookcode 中创建 entity 子包,并在包 com.bookcode.entity 中创建类 BeanInjector,代码如例 7-2 所示。

【例 7-2】 类 BeanInjector 的代码示例。

```
package com.bookcode.entity;
import org.springframework.context.annotation.Bean;
import org.springframework.stereotype.Component;
```

```java
import org.springframework.util.StringUtils;
import java.util.UUID;
@Component
public class BeanInjector {
    private BlockChain blockChain;
    private String nodeId;
    @Bean("nodeId")
    private String nodeId() {
        synchronized (this) {
            if (StringUtils.isEmpty(nodeId)) {
                nodeId = UUID.randomUUID().toString().replace("-", "");
            }
        }
        return nodeId;
    }
}
```

7.1.3 创建类 BlockChain

在包 com.bookcode.entity 中创建类 BlockChain，代码如例 7-3 所示。该类实现了 PoW 算法。

【例 7-3】 类 BlockChain 的代码示例。

```java
package com.bookcode.entity;
import com.alibaba.fastjson.TypeReference;
import com.alibaba.fastjson.annotation.JSONField;
import com.bookcode.util.FastJsonUtil;
import com.bookcode.util.SHAUtils;
import com.fasterxml.jackson.annotation.JsonIgnore;
import io.swagger.annotations.ApiModel;
import io.swagger.annotations.ApiModelProperty;
import lombok.Data;
import org.apache.commons.collections4.MapUtils;
import org.springframework.util.CollectionUtils;
import org.springframework.web.client.RestTemplate;
import java.math.BigDecimal;
import java.util.*;
@Data
@ApiModel(description = "区块链")
public class BlockChain {
    private final static FastJsonUtil jsonUtil = new FastJsonUtil();
    @ApiModelProperty(value = "当前交易列表", dataType = "List<Transaction>")
    @JSONField(serialize = false)
    @JsonIgnore
    private List<Transaction> currentTransactions;
    @ApiModelProperty(value = "所有交易列表", dataType = "List<Transaction>")
    private List<Transaction> transactions;
```

```java
    @ApiModelProperty(value = "区块列表", dataType = "List<BlockChain>")
    @JSONField(serialize = false)
    @JsonIgnore
    private List<BlockChain> chain;
    @ApiModelProperty(value = "集群的节点列表", dataType = "Set<String>")
    @JSONField(serialize = false)
    @JsonIgnore
    private Set<String> nodes;
    @ApiModelProperty(value = "上一个区块的哈希值", dataType = "String", example = "f461ac428043f328309da7cac33803206cea9912f0d4e8d8cf2786d21e5ff403")
    private String previousHash = "";
    @ApiModelProperty(value = "工作量证明", dataType = "Integer", example = "100")
    private Integer proof = 0;
    @ApiModelProperty(value = "当前区块的索引序号", dataType = "Long", example = "2")
    private Long index = 0L;
    @ApiModelProperty(value = "当前区块的时间戳", dataType = "Long", example = "1526458171000")
    private Long timestamp = 0L;
    @ApiModelProperty(value = "当前区块的哈希值", dataType = "String", example = "g451ac428043f328309da7cac33803206cea9912f0d4e8d8cf2786d21e5ff401")
    private String hash;
    public BlockChain() {
        currentTransactions = new ArrayList<>();
        chain = new ArrayList<>();
        transactions = new ArrayList<>();
        nodes = new HashSet<>();
    }
    public String getHash() {
        String json = jsonUtil.toJson(this.getCurrentTransactions()) +
                jsonUtil.toJson(this.getTransactions()) +
                jsonUtil.toJson(this.getChain()) +
                   this.getPreviousHash() + this.getProof() + this.getIndex() + this.getTimestamp();
        hash = SHAUtils.getSHA256Str(json);
        return hash;
    }
    @JSONField(serialize = false)
    @JsonIgnore
    public BlockChain getLastBlock() {
        if (CollectionUtils.isEmpty(chain)) {
            return null;
        }
        return chain.get(chain.size() - 1);
    }
    public void registerNode(String address) {
        nodes.add(address);
    }
    public BlockChain newBlock(Integer proof, String previousHash) {
        BlockChain block = new BlockChain();
        block.index = chain.size() + 1L;
```

```java
            block.timestamp = System.currentTimeMillis();
            block.transactions.addAll(currentTransactions);
            block.proof = proof;
            block.previousHash = previousHash;
            currentTransactions.clear();
            chain.add(block);
            return block;
        }
        public Long newTransaction(String sender, String recipient, BigDecimal amount) {
            Transaction transaction = new Transaction();
            transaction.setSender(sender);
            transaction.setRecepient(recipient);
            transaction.setAmount(amount);
            currentTransactions.add(transaction);
            return getLastBlock().index + 1;
        }
        public Integer proofOfWork(Integer lastProof) {
            int proof = 0;
            while (!validProof(lastProof, proof)) {
                proof += 1;
            }
            return proof;
        }
        public Boolean validProof(Integer lastProof, Integer proof) {
            System.out.println("挖矿的工作量(PoW)证明如下所示:");
            System.out.println("有效工作证明(PoW)==>结束证明:" + lastProof + ",证明:" + proof);
            String guessHash = SHAUtils.getSHA256Str(String.format("{%d}{%d}", lastProof, proof));
            return guessHash.startsWith("00");
        }
        public void newSeedBlock() {
            newBlock(100, "1");
        }
        public boolean validChain(List<BlockChain> chain) {
            if (CollectionUtils.isEmpty(chain)) {
                return false;
            }
            BlockChain previousBlock = chain.get(0);
            int currentIndex = 1;
            while (currentIndex < chain.size()) {
                BlockChain block = chain.get(currentIndex);
                if (!block.getPreviousHash().equals(previousBlock.getHash())) {
                    return false;
                }
                if (!validProof(previousBlock.getProof(), block.getProof())) {
                    return false;
                }
                previousBlock = block;
                currentIndex += 1;
```

```java
            }
            return true;
        }
    public boolean resolveConflicts() {
        int maxLength = getChain().size();
        List<BlockChain> newChain = new ArrayList<>();
        for (String node : getNodes()) {
            RestTemplate template = new RestTemplate();
            Map map = template.getForObject(node + "chain", Map.class);
            int length = MapUtils.getInteger(map, "length");
            String json = jsonUtil.toJson(MapUtils.getObject(map, "chain"));
            List<BlockChain> chain = jsonUtil.fromJson(json, new TypeReference<List<BlockChain>>() {
            });
            if (length > maxLength && validChain(chain)) {
                maxLength = length;
                newChain = chain;
            }
        }
        if (!CollectionUtils.isEmpty(newChain)) {
            this.chain = newChain;
            return true;
        }
        return false;
    }
}
```

7.1.4 创建类 RegisterRequest

在包 com.bookcode.entity 中创建类 RegisterRequest，代码如例 7-4 所示。

【例 7-4】 类 RegisterRequest 的代码示例。

```java
package com.bookcode.entity;
import io.swagger.annotations.ApiModel;
import io.swagger.annotations.ApiModelProperty;
import lombok.Data;
import java.util.List;
@Data
@ApiModel(description = "注册节点-请求参数")
public class RegisterRequest {
    @ApiModelProperty(value = "集群中的节点列表", dataType = "List<String>")
    private List<String> nodes;
}
```

7.1.5 创建类 Transaction

在包 com.bookcode.entity 中创建类 Transaction,代码如例 7-5 所示。

【例 7-5】 类 Transaction 的代码示例。

```java
package com.bookcode.entity;
import io.swagger.annotations.ApiModel;
import io.swagger.annotations.ApiModelProperty;
import lombok.Data;
import java.math.BigDecimal;
@Data
@ApiModel(description = "交易信息")
public class Transaction {
    @ApiModelProperty(value = "交易发起方", dataType = "String", example = "赵高")
    private String sender;
    @ApiModelProperty(value = "交易接收方", dataType = "String", example = "李斯")
    private String recepient;
    @ApiModelProperty(value = "交易额", dataType = "BigDecimal", example = "1.11")
    private BigDecimal amount;
}
```

7.1.6 创建类 FastJsonUtil

在包 com.bookcode 中创建 util 子包,并在包 com.bookcode.util 中创建类 FastJsonUtil,代码如例 7-6 所示。

【例 7-6】 类 FastJsonUtil 的代码示例。

```java
package com.bookcode.util;
import com.alibaba.fastjson.JSON;
import com.alibaba.fastjson.TypeReference;
import com.alibaba.fastjson.serializer.JSONLibDataFormatSerializer;
import com.alibaba.fastjson.serializer.SerializeConfig;
import com.alibaba.fastjson.serializer.SerializerFeature;
public class FastJsonUtil {
    private static final SerializeConfig config;
    static {
        config = new SerializeConfig();
        config.put(java.util.Date.class, new JSONLibDataFormatSerializer());
//使用和 json-lib 兼容的日期输出格式
        config.put(java.sql.Date.class, new JSONLibDataFormatSerializer());
    }
        private static final SerializerFeature [ ] features = { SerializerFeature.
WriteMapNullValue, SerializerFeature.WriteNullListAsEmpty,
            SerializerFeature.WriteNullNumberAsZero,
            SerializerFeature.WriteNullBooleanAsFalse,
```

```java
            SerializerFeature.WriteNullStringAsEmpty
    };
    public static String toJson(Object object) {
        if (null == object) {
            return null;
        }
        return JSON.toJSONString(object, config, features);
    }
    public <T> T fromJson(String jsonString, TypeReference<T> typeReference) {
        try {
            return JSON.parseObject(jsonString, typeReference);
        } catch (Exception e) {
            e.printStackTrace();
            throw e;
        }
    }
}
```

7.1.7 创建类 SHAUtils

在包 com.bookcode.util 中创建类 SHAUtils,代码如例 7-7 所示。

【例 7-7】 类 SHAUtils 的代码示例。

```java
package com.bookcode.util;
import java.io.UnsupportedEncodingException;
import java.security.MessageDigest;
import java.security.NoSuchAlgorithmException;
import org.apache.commons.codec.binary.Hex;
public class SHAUtils {
    public static String getSHA256Str(String str) {
        MessageDigest messageDigest;
        String encdeStr = "";
        try {
            messageDigest = MessageDigest.getInstance("SHA-256");
            byte[] hash = messageDigest.digest(str.getBytes("UTF-8"));
            encdeStr = Hex.encodeHexString(hash);
        } catch (NoSuchAlgorithmException e) {
            e.printStackTrace();
        } catch (UnsupportedEncodingException e) {
            e.printStackTrace();
        }
        return encdeStr;
    }
}
```

7.1.8 创建类 SwaggerConfig

在包 com.bookcode 中创建 config 子包,并在包 com.bookcode.config 中创建类 SwaggerConfig,代码如例 7-8 所示。

【例 7-8】 类 SwaggerConfig 的代码示例。

```
package com.bookcode.config;
import org.springframework.context.annotation.Bean;
import org.springframework.context.annotation.Configuration;
import org.springframework.web.servlet.config.annotation.ResourceHandlerRegistry;
import org.springframework.web.servlet.config.annotation.WebMvcConfigurer;
import org.springframework.web.servlet.config.annotation.WebMvcConfigurerAdapter;
import springfox.documentation.builders.ApiInfoBuilder;
import springfox.documentation.builders.PathSelectors;
import springfox.documentation.builders.RequestHandlerSelectors;
import springfox.documentation.service.ApiInfo;
import springfox.documentation.service.Contact;
import springfox.documentation.spi.DocumentationType;
import springfox.documentation.spring.web.plugins.Docket;
import springfox.documentation.swagger2.annotations.EnableSwagger2;
@Configuration
@EnableSwagger2
public class SwaggerConfig {
    @Bean
    public Docket createRestApi() {
        return new Docket(DocumentationType.SWAGGER_2)
                .apiInfo(apiInfo())
                .select()
                .apis(RequestHandlerSelectors.basePackage("com.bookcode.controller"))
                .paths(PathSelectors.any())
                .build();
    }
    @Bean
    public WebMvcConfigurer addResourceHandlers() {
        return new WebMvcConfigurerAdapter() {
            @Override
            public void addResourceHandlers(ResourceHandlerRegistry registry) {
                registry.addResourceHandler("swagger-ui.html")
                        .addResourceLocations("classpath:/META-INF/resources/");
                registry.addResourceHandler("/webjars/**")
                        .addResourceLocations("classpath:/META-INF/resources/webjars/");
            }
        };
    }
    private ApiInfo apiInfo() {
```

```
            return new ApiInfoBuilder()
                    .title("Spring Boot 区块链在线 API 文档")
                    .description("区块链示例")
                .contact(new Contact("文档官方网站", "http://localhost:8080/index.html",
"163@126.com"))
                    .version("1.0.0")
                    .build();
    }
}
```

7.1.9　创建类 BlockChainController

在包 com.bookcode 中创建 controller 子包,并在包 com.bookcode.controller 中创建类 BlockChainController,代码如例 7-9 所示。

【例 7-9】　类 BlockChainController 的代码示例。

```
package com.bookcode.controller;
import com.bookcode.entity.BlockChain;
import com.bookcode.entity.RegisterRequest;
import com.bookcode.entity.Transaction;
import io.swagger.annotations.Api;
import io.swagger.annotations.ApiOperation;
import org.apache.commons.collections4.CollectionUtils;
import org.springframework.context.annotation.Scope;
import org.springframework.stereotype.Controller;
import org.springframework.web.bind.annotation.*;
import javax.annotation.Resource;
import java.math.BigDecimal;
import java.util.HashMap;
import java.util.Map;
@Scope("prototype")
@Controller
@RequestMapping(value = "/")
@Api(consumes = "application/json",
        produces = "application/json",
        protocols = "http",
        basePath = "/", value = "区块链示例")
public class BlockChainController {
    @Resource
    private BlockChain blockChain;
    @Resource
    private String nodeId;
    @RequestMapping("/")
    public String home() {
        return "redirect:/swagger-ui.html";
```

```java
}
@GetMapping(value = "/chain")
@ResponseBody
@ApiOperation(value = "查看完整的区块链")
public Map<String, Object> fullChain() {
    Map<String, Object> map = new HashMap<>();
    map.put("chain", blockChain.getChain());
    map.put("length", blockChain.getChain().size());
    return map;
}
@PostMapping(value = "/transactions/new")
@ResponseBody
@ApiOperation(value = "创建新交易")
public Map<String, Object> newTransaction(@RequestBody Transaction transaction) {
    long index = blockChain.newTransaction(transaction.getSender(),
            transaction.getRecepient(),
            transaction.getAmount());
    Map<String, Object> map = new HashMap<>();
    map.put("message", String.format("交易信息被新增到第 %d 块区块.", index));
    return map;
}
@GetMapping(value = "/mine")
@ResponseBody
@ApiOperation(value = "挖矿")
public Map<String, Object> mine() {
    Map<String, Object> map = new HashMap<>();
    BlockChain lastBlock = blockChain.getLastBlock();
    Integer lastProof = lastBlock.getProof();
    Integer proof = blockChain.proofOfWork(lastProof);
    blockChain.newTransaction("0", nodeId, BigDecimal.ONE);
    BlockChain block = blockChain.newBlock(proof, lastBlock.getHash());
    map.put("message", "New Block Forged");
    map.put("index", block.getIndex());
    map.put("transactions", block.getTransactions());
    map.put("proof", block.getProof());
    map.put("previousHash", block.getPreviousHash());
    return map;
}
@PostMapping(value = "/register")
@ResponseBody
@ApiOperation(value = "注册集群节点")
public Map<String, Object> register(@RequestBody RegisterRequest request) {
    if (CollectionUtils.isNotEmpty(request.getNodes())) {
        for (String n : request.getNodes()) {
            blockChain.registerNode(n);
        }
    }
    Map<String, Object> map = new HashMap<>();
```

```java
            map.put("message", "增加新节点.");
            map.put("totalNodes", blockChain.getNodes());
            return map;
        }
        @GetMapping(value = "/resolve")
        @ResponseBody
        @ApiOperation(value = "解决不同节点间的数据冲突")
        public Map<String, Object> resolve() {
            boolean replaced = blockChain.resolveConflicts();
            Map<String, Object> map = new HashMap<>();
            if (replaced) {
                map.put("message", "因为冲突,我们的区块链被替换了.");
                map.put("newChain", blockChain.getChain());
            } else {
                map.put("message", "我们的区块链和其他区块链不存在冲突,可以提交.");
                map.put("chain", blockChain.getChain());
            }
            return map;
        }
        @GetMapping(value = "/validate")
        @ResponseBody
        @ApiOperation(value = "验证区块链自身是否合法")
        public Map<String, Object> validate() {
            boolean result = blockChain.validChain(blockChain.getChain());
            Map<String, Object> map = new HashMap<>();
            if (result) {
                map.put("message", "该区块链合法.");
            } else {
                map.put("message", "该区块链不合法.");
            }
            return map;
        }
    }
```

7.1.10　修改配置文件 application.properties

修改配置文件 application.properties，修改后的代码如例 7-10 所示。

【例 7-10】　修改后的配置文件 application.properties 代码示例。

```
swagger.enabled = true
swagger.title = Spring Boot 区块链在线 API 文档
swagger.description = 区块链示例
swagger.contact.url = http://localhost:8080/index.html
swagger.contact.email = 123@163.com
swagger.base-package = com.bookcode
```

```
swagger.base-path=/**
swagger.exclude-path=/error,/adminlte/**,/actuator
```

7.1.11 创建文件 index.html

在目录 src/main/resources/static 下,创建文件 index.html,代码如例 7-11 所示。

【例 7-11】 文件 index.html 的代码示例。

```
<!DOCTYPE html>
<html lang="en">
<head>
    <meta charset="UTF-8">
    <title>区块链</title>
</head>
<body>
    <H1>区块链相关文档</H1>
</body>
</html>
```

7.1.12 运行程序

运行程序后,在浏览器中输入 localhost:8080,并单击链接 Models 后的结果如图 7-1 所示。单击图 7-1 中 block-chain-controller 链接后的结果如图 7-2 所示。单击图 7-1 中 http://localhost:8080/v2/api-docs 链接后的结果如图 7-3 所示。单击图 7-1 中"文档官方网站-Website"链接后的结果如图 7-4 所示。单击图 7-1 中"Send email to 文档官方网站"链接后的结果如图 7-5 所示。

图 7-1 在浏览器中输入 localhost:8080 并单击链接 Models 后的结果

第7章　区块链应用中共识算法的实现

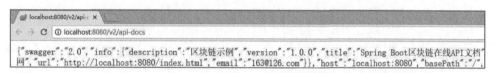

图 7-2　单击图 7-1 中 block-chain-controller 链接后的结果

图 7-3　单击图 7-1 中 http://localhost:8080/v2/api-docs 链接后的结果

图 7-4　单击图 7-1 中"文档官方网站　　图 7-5　单击图 7-1 中"Send email to 文档
　　　　-Website"链接后的结果　　　　　　　　　官方网站"链接后的结果

7.2 PBFT算法的实现

视频讲解

PBFT算法解决了原始拜占庭容错算法效率不高的问题,将算法复杂度由指数级降低到多项式级,使得拜占庭容错算法可以在实际系统中应用。PBET算法根据在分布式网络中节点间互相交换信息后选择大多数的结果作为共识机制形成办法,一个节点代表一票。PBET算法将容错量控制在全部节点数的1/3,即只要有超过2/3节点正常整个系统便可正常运作。

7.2.1 创建项目并添加依赖

用IDEA创建完项目poaexample之后,确保在文件pom.xml中<dependencies>和</dependencies>之间添加了Guava、Lombok等依赖,代码如例7-12所示。

【例7-12】 添加Guava、Lombok等依赖的代码示例。

```xml
<dependency>
    <groupId>com.google.guava</groupId>
    <artifactId>guava</artifactId>
    <version>24.0-jre</version>
</dependency>
<dependency>
    <groupId>org.projectlombok</groupId>
    <artifactId>lombok</artifactId>
    <version>1.18.4</version>
</dependency>
<dependency>
    <groupId>org.apache.commons</groupId>
    <artifactId>commons-lang3</artifactId>
    <version>3.8.1</version>
</dependency>
<dependency>
    <groupId>org.springframework.boot</groupId>
    <artifactId>spring-boot-starter-web</artifactId>
</dependency>
```

7.2.2 创建类PbftMsg

在包com.bookcode中创建entity子包,并在包com.bookcode.entity中创建类PbftMsg,代码如例7-13所示。

【例7-13】 类PbftMsg的代码示例。

```java
package com.bookcode.entity;
import lombok.Data;
```

```java
@Data
public class PbftMsg {
    private boolean isOk;
    private int type;
    private int node;
    private int onode;
    private int vnum;
    private int no;
    private long time;
    private String data;
    public PbftMsg(int type,int node) {
        this.type = type;
        this.node = node;
        this.onode = node;
        this.time = System.currentTimeMillis();
        this.isOk = true;
    }
    public PbftMsg(PbftMsg msg) {
        this.isOk = msg.isOk;
        this.type = msg.type;
        this.node = msg.node;
        this.onode = msg.onode;
        this.vnum = msg.vnum;
        this.data = msg.data;
        this.time = msg.time;
        this.no = msg.no;
    }
    public String getKey(){
        return getDataKey() + "|@|" + getNode();
    }
    public String getDataKey(){
        return getData() + "|@|" + getNo();
    }
    @Override
    public String toString() {
return "PbftMsg [isOk = " + isOk + ", type = " + type + ", node = " + node + ", vnum = " +
 vnum + ", time = " + time + ", data = " + data + ", no = " + no + "]" + ", onode = " + onode
 + "]";
    }
}
```

7.2.3 创建类 Pbft

在包 com.bookcode 中创建 util 子包,并在包 com.bookcode.util 中创建类 Pbft,代码如例 7-14 所示。

【例 7-14】 类 Pbft 的代码示例。

```java
package com.bookcode.util;
import java.util.List;
import java.util.Map;
import java.util.Map.Entry;
import java.util.Set;
import java.util.Timer;
import java.util.TimerTask;
import java.util.concurrent.BlockingQueue;
import java.util.concurrent.atomic.AtomicInteger;
import java.util.concurrent.atomic.AtomicLong;
import com.bookcode.entity.PbftMsg;
import org.apache.commons.lang3.StringUtils;
import com.google.common.collect.Lists;
import com.google.common.collect.Maps;
import com.google.common.collect.Queues;
import com.google.common.collect.Sets;
import com.google.common.util.concurrent.AtomicLongMap;
public class Pbft {
    public int size;
    public int maxf;
    public static final int CV = -2;
    public static final int VIEW = -1;
    public static final int REQ = 0;
    public static final int PP = 1;
    public static final int PA = 2;
    public static final int CM = 3;
    public static final int REPLY = 4;
    private volatile boolean isRun = false;
    private BlockingQueue<PbftMsg> qbm = Queues.newLinkedBlockingQueue();
    private Set<String> votes_pre = Sets.newConcurrentHashSet();
    private Set<String> votes_pare = Sets.newConcurrentHashSet();
    private AtomicLongMap<String> aggre_pare = AtomicLongMap.create();
    private Set<String> votes_comm = Sets.newConcurrentHashSet();
    private AtomicLongMap<String> aggre_comm = AtomicLongMap.create();
    private Map<String,PbftMsg> doneMsg = Maps.newConcurrentMap();
    private Map<String,PbftMsg> applyMsg = Maps.newConcurrentMap();
    private int index;
    private int view;
    private volatile boolean viewOk;
    private AtomicLongMap<Integer> vnumAggreCount = AtomicLongMap.create();
    private Set<String> votes_vnum = Sets.newConcurrentHashSet();
    private AtomicLong replyCount = new AtomicLong();
    private Map<String,Long> timeOuts = Maps.newHashMap();
    private Map<String,Long> timeOutsReq = Maps.newHashMap();
    private BlockingQueue<PbftMsg> reqQueue = Queues.newLinkedBlockingDeque(100);
    private PbftMsg curMsg;
    private volatile AtomicInteger genNo = new AtomicInteger(0);
    private Timer timer;
```

```java
public Pbft(int node,int size) {
    this.index = node;
    this.size = size;
    this.maxf = (size-1)/3;
    timer = new Timer("timer" + node);
}
public Pbft start(){
    new Thread(new Runnable() {
        @Override
        public void run() {
            while (true) {
                try {
                    PbftMsg msg = qbm.take();
                    doAction(msg);
                } catch (InterruptedException e) {
                    e.printStackTrace();
                }
            }
        }
    }).start();
    isRun = true;
    timer.schedule(new TimerTask() {
        int co = 0;
        @Override
        public void run() {
            if(co == 0){
                pubView();
            }
            co++;
            doReq();
            checkTimer();
        }
    }, 10, 100);
    return this;
}
private void checkTimer() {
    List<String> remo = Lists.newArrayList();
    for(Entry<String, Long> item : timeOuts.entrySet()){
        if(System.currentTimeMillis() - item.getValue() > 300){
            System.out.println("投票无效[" + index + "]:" + item.getKey());
            remo.add(item.getKey());
        }
    }
    remo.forEach((it) ->{
        remove(it);
    });
    remo.clear();
    for(Entry<String, Long> item : timeOutsReq.entrySet()){
        if(System.currentTimeMillis() - item.getValue() > 600){
            remo.add(item.getKey());
```

```java
            }
        }
        remo.forEach((data) ->{
            System.out.println("请求主节点超时[" + index + "]:" + data);
            timeOutsReq.remove(data);
            if(curMsg != null && curMsg.getData().equals(data)){
                vnumAggreCount.incrementAndGet(this.view + 1);
                votes_vnum.add(index + "@" + (this.view + 1));
                PoAUtil.publish(curMsg);
            }else{
                if(!this.viewOk) return;
                this.viewOk = false;
                PbftMsg cv = new PbftMsg(CV, this.index);
                cv.setVnum(this.view + 1);
                PoAUtil.publish(cv);
            }
        });
    }
    public void pubView(){
        PbftMsg sed = new PbftMsg(VIEW, index);
        PoAUtil.publish(sed);
    }
    public void req(String data) throws InterruptedException{
        PbftMsg req = new PbftMsg(Pbft.REQ, this.index);
        req.setData(data);
        reqQueue.put(req);
    }
    protected boolean doReq() {
        if(!viewOk || curMsg != null)return false; //上一个请求还没发完/视图初始化中
        curMsg = reqQueue.poll();
        if(curMsg == null)return false;
        curMsg.setVnum(this.view);
        doSendCurMsg();
        return true;
    }
    void doSendCurMsg(){
        timeOutsReq.put(curMsg.getData(), System.currentTimeMillis());
        PoAUtil.send(getPriNode(view), curMsg);
    }
    private void remove(String it) {
        votes_pre.remove(it);
        votes_pare.removeIf((vp) ->{
            return StringUtils.startsWith(vp, it);
        });
        votes_comm.removeIf((vp) ->{
            return StringUtils.startsWith(vp, it);
        });
        aggre_pare.remove(it);
        aggre_comm.remove(it);
        timeOuts.remove(it);
```

```java
    }
    protected boolean doAction(PbftMsg msg) {
        if(!isRun) return false;
        if(msg != null){
            System.out.println("收到消息[" + index + "]:" + msg);
            switch (msg.getType()) {
                case PP:
                    onPrePrepare(msg);
                    break;
                case PA:
                    onPrepare(msg);
                    break;
                case CM:
                    onCommit(msg);
                    break;
                case REPLY:
                    onReply(msg);
                    break;
                case VIEW:
                    onGetView(msg);
                    break;
                case REQ:
                    onReq(msg);
                    break;
                case CV:
                    onChangeView(msg);
                    break;
                default:
                    break;
            }
            return true;
        }
        return false;
    }
    private void onChangeView(PbftMsg msg) {
        String vkey = msg.getNode() + "@" + msg.getVnum();
        if(votes_vnum.contains(vkey)){
            return;
        }
        votes_vnum.add(vkey);
        long count = vnumAggreCount.incrementAndGet(msg.getVnum());
        if(count >= 2 * maxf + 1){
            vnumAggreCount.clear();
            this.view = msg.getVnum();
            viewOk = true;
            System.out.println("视图变更完成[" + index + "]: " + view);
            if(curMsg != null){
                curMsg.setVnum(this.view);
                System.out.println("请求重传[" + index + "]: " + curMsg);
                doSendCurMsg();
```

```java
            }
        }
    }
    private void onReq(PbftMsg msg) {
        if(!msg.isOk()) return;
        PbftMsg sed = new PbftMsg(msg);
        sed.setNode(index);
        if(msg.getVnum() < view) return;
        if(msg.getVnum() == index){
            if(applyMsg.containsKey(msg.getDataKey())) return;
            applyMsg.put(msg.getDataKey(), msg);
            sed.setType(PP);
            int no = genNo.incrementAndGet();
            sed.setNo(no);
            PoAUtil.publish(sed);
        }else if(msg.getNode() != index){
            if(doneMsg.containsKey(msg.getDataKey())){
                sed.setType(REPLY);
                PoAUtil.send(msg.getNode(), sed);
            }else{
                votes_vnum.add(msg.getNode() + "@" + (msg.getVnum() + 1));
                vnumAggreCount.incrementAndGet(msg.getVnum() + 1);
                System.out.println("转发主节点[" + index + "]:" + msg);
                PoAUtil.send(getPriNode(view), sed);
                timeOutsReq.put(msg.getData(), System.currentTimeMillis());
            }
        }
    }
    private void onReply(PbftMsg msg) {
        if(curMsg == null || !curMsg.getData().equals(msg.getData()))return;
        long count = replyCount.incrementAndGet();
        replyCount.set(0);
        curMsg = null;
        doSomething(msg);
    }
    private void doSomething(PbftMsg msg) {
        System.out.println("请求执行成功[" + index + "]:" + msg);
    }
    private void onGetView(PbftMsg msg) {
        if(msg.getData() == null){
            PbftMsg sed = new PbftMsg(msg);
            sed.setNode(index);
            sed.setVnum(view);
            sed.setData("initview");
            PoAUtil.send(msg.getNode(), sed);
        }else{
            if(this.viewOk)return;
            long count = vnumAggreCount.incrementAndGet(msg.getVnum());
            if(count >= 2 * maxf + 1){
                vnumAggreCount.clear();
```

```java
                this.view = msg.getVnum();
                viewOk = true;
                System.out.println("视图初始化完成[" + index + "]: " + view);
            }
        }
    }
    private void onCommit(PbftMsg msg) {
        if(!checkMsg(msg,false)) return;
        String key = msg.getKey();
        if(votes_comm.contains(key)){
            return;
        }
        if(!votes_pare.contains(key)){
            return;
        }
        votes_comm.add(key);
        long agCou = aggre_comm.incrementAndGet(msg.getDataKey());
        if(agCou >= 2 * maxf + 1){
            remove(msg.getDataKey());
            if(msg.getNode() != index){
                this.genNo.set(msg.getNo());
            }
            if(msg.getOnode() == index){
                onReply(msg);
            }else{
                PbftMsg sed = new PbftMsg(msg);
                sed.setType(REPLY);
                sed.setNode(index);
                doSomething(sed);
            }
        }
    }
    private void onPrepare(PbftMsg msg) {
        if(!checkMsg(msg,false)) {
            System.out.println("异常消息[" + index + "]:" + msg);
            return;
        }
        String key = msg.getKey();
        if(votes_pare.contains(key)){
            return;
        }
        if(!votes_pre.contains(msg.getDataKey())){
            return;
        }
        votes_pare.add(key);
        long agCou = aggre_pare.incrementAndGet(msg.getDataKey());
        if(agCou >= 2 * maxf + 1){
            aggre_pare.remove(msg.getDataKey());
            PbftMsg sed = new PbftMsg(msg);
            sed.setType(CM);
```

```java
            sed.setNode(index);
            doneMsg.put(sed.getDataKey(), sed);
            PoAUtil.publish(sed);
        }
    }
    private void onPrePrepare(PbftMsg msg) {
        if(!checkMsg(msg,true)) return;
        String key = msg.getDataKey();
        if(votes_pre.contains(key)){
            return;
        }
        votes_pre.add(key);
        timeOuts.put(key, System.currentTimeMillis());
        timeOutsReq.remove(msg.getData());
        PbftMsg sed = new PbftMsg(msg);
        sed.setType(PA);
        sed.setNode(index);
        PoAUtil.publish(sed);
    }
    public boolean checkMsg(PbftMsg msg,boolean isPre){
        return (msg.isOk() && msg.getVnum() == view
                && (!isPre || msg.getNode() == index || (getPriNode(view) == msg.getNode()
 && msg.getNo() > genNo.get())));
    }
    public int getPriNode(int view2){
        return view2 % size;
    }
    public void push(PbftMsg msg){
        try {
            this.qbm.put(msg);
        } catch (InterruptedException e) {
            e.printStackTrace();
        }
    }
    public int getView() {
        return view;
    }
    public void setView(int view) {
        this.view = view;
    }
    public void close(){
        System.out.println("宕机[" + index + "]---------------");
        this.isRun = false;
    }
    public int getIndex(){
        return this.index;
    }
    public void back() {
        System.out.println("恢复[" + index + "]---------------");
        this.isRun = true;
```

        }
    }

7.2.4 创建类 PoAUtil

在包 com.bookcode.util 中创建类 PoAUtil，代码如例 7-15 所示。

【例 7-15】 类 PoAUtil 的代码示例。

```java
package com.bookcode.util;
import com.bookcode.entity.PbftMsg;
import com.google.common.collect.Lists;
import lombok.Data;
import java.util.List;
@Data
public class PoAUtil {
    public static List<Pbft> nodes = Lists.newArrayList();
    public static long[] net = new long[99];
    public static void publish(PbftMsg msg){
        for(Pbft pbft:nodes){
            TimerManager.schedule(()->{
                pbft.push(new PbftMsg(msg));
                return null;
            }, net[msg.getNode() * 10 + pbft.getIndex()]);
        }
    }
    public static void send(int toIndex,PbftMsg msg){
        TimerManager.schedule(()->{
            nodes.get(toIndex).push(msg);
            return null;
        }, net[msg.getNode() * 10 + toIndex]);
    }
}
```

7.2.5 创建类 TimerManager

在包 com.bookcode.util 中创建类 TimerManager，代码如例 7-16 所示。

【例 7-16】 类 TimerManager 的代码示例。

```java
package com.bookcode.util;
import java.util.concurrent.Executors;
import java.util.concurrent.ScheduledExecutorService;
import java.util.concurrent.TimeUnit;
import java.util.function.Supplier;
public class TimerManager {
```

```java
        private volatile static ScheduledExecutorService executorService = Executors.
newScheduledThreadPool(3);
    public static void schedule(Supplier<?> action, long delay){
        executorService.schedule(new Runnable() {
            @Override
            public void run() {
                action.get();
            }
        }, delay, TimeUnit.MILLISECONDS);
    }
    public static void scheduleAtFixedRate(Supplier<?> action, long initialDelay, long period ){
        executorService.scheduleAtFixedRate(new Runnable() {
            @Override
            public void run() {
                action.get();
            }
        }, initialDelay,period, TimeUnit.MILLISECONDS);
    }

    public static void scheduleWithFixedDelay(Supplier<?> action,long initialDelay, long period ){
        executorService.scheduleWithFixedDelay(new Runnable() {
            @Override
            public void run() {
                action.get();
            }
        }, initialDelay,period, TimeUnit.MILLISECONDS);
    }
}
```

7.2.6 创建类 PbftController

在包 com.bookcode 中创建 controller 子包,并在包 com.bookcode.controller 中创建类 PbftController,代码如例 7-17 所示。

【例 7-17】 类 PbftController 的代码示例。

```java
package com.bookcode.controller;
import com.bookcode.util.Pbft;
import com.bookcode.util.PoAUtil;
import org.apache.commons.lang3.RandomUtils;
import org.springframework.web.bind.annotation.GetMapping;
import org.springframework.web.bind.annotation.RestController;
import java.util.Random;
@RestController
public class PbftController {
```

```
public static final int size = 4;
private static Random r = new Random();
@GetMapping("/testPbft")
public String pbft() throws InterruptedException {
    for(int i = 0;i < size;i++){
        PoAUtil.nodes.add(new Pbft(i,4).start());
    }
    for(int i = 0;i < size;i++){
        for(int j = 0;i < size;i++){
            if(i != j){
                PoAUtil.net[i * 10 + j] = RandomUtils.nextLong(10, 60);
            }else{
                PoAUtil.net[i * 10 + j] = 10;
            }
        }
    }
    for(int i = 0;i < 1;i++){
        int node = r.nextInt(size);
        PoAUtil.nodes.get(node).req("test" + i);
    }
    Thread.sleep(1000);
    PoAUtil.nodes.get(0).close();
    for(int i = 2;i < 4;i++){
        PoAUtil.nodes.get(i).req("testD" + i);
    }
    Thread.sleep(1000);
    PoAUtil.nodes.get(0).back();
    for(int i = 1;i < 2;i++){
        PoAUtil.nodes.get(i).req("testB" + i);
    }
    return "Tested Pbft.";
}
```

7.2.7 运行程序

运行程序后,在浏览器中输入 localhost:8080/testPbft,浏览器中的结果如图 7-6 所示。在控制台中的输出结果如例 7-18 所示,其中省略号是"收到消息"的具体内容(格式和第一行收到消息一样,内容不完全相同)。为了节省空间和更好地看清整个输出情况,省略了部分收到消息的内容。

图 7-6 在浏览器中输入 localhost:8080/testPbft 后浏览器中显示的结果

【例 7-18】 控制台中输出的结果示例。

```
收到消息[1]:PbftMsg [isOk = true, type = -1, node = 0, vnum = 0, time = 1551339456178, data = null,
no = 0], onode = 0]
...
视图初始化完成[1]: 0
收到消息[3]:PbftMsg [isOk = true, type = -1, node = 1, vnum = 0, time = 1551339456179, data = initview, no = 0], onode = 3]
视图初始化完成[3]: 0
视图初始化完成[2]: 0
收到消息[0]:PbftMsg [isOk = true, type = -1, node = 0, vnum = 0, time = 1551339456178, data = null, no = 0], onode = 0]
...
视图初始化完成[0]: 0
收到消息[0]:PbftMsg [isOk = true, type = -1, node = 2, vnum = 0, time = 1551339456179, data = null, no = 0], onode = 2]
...
请求执行成功[2]:PbftMsg [isOk = true, type = 4, node = 2, vnum = 0, time = 1551339456176, data = test0, no = 1], onode = 0]
请求执行成功[3]:PbftMsg [isOk = true, type = 4, node = 3, vnum = 0, time = 1551339456176, data = test0, no = 1], onode = 0]
请求执行成功[1]:PbftMsg [isOk = true, type = 4, node = 1, vnum = 0, time = 1551339456176, data = test0, no = 1], onode = 0]
收到消息[0]:PbftMsg [isOk = true, type = 1, node = 0, vnum = 0, time = 1551339456176, data = test0, no = 1], onode = 0]
...
请求执行成功[0]:PbftMsg [isOk = true, type = 3, node = 2, vnum = 0, time = 1551339456176, data = test0, no = 1], onode = 0]
收到消息[0]:PbftMsg [isOk = true, type = 3, node = 0, vnum = 0, time = 1551339456176, data = test0, no = 1], onode = 0]
宕机[0]----------------
请求主节点超时[3]:testD3
请求主节点超时[2]:testD2
收到消息[1]:PbftMsg [isOk = true, type = 0, node = 2, vnum = 0, time = 1551339457177, data = testD2, no = 0], onode = 2]
转发主节点[1]:PbftMsg [isOk = true, type = 0, node = 2, vnum = 0, time = 1551339457177, data = testD2, no = 0], onode = 2]
收到消息[2]:PbftMsg [isOk = true, type = 0, node = 2, vnum = 0, time = 1551339457177, data = testD2, no = 0], onode = 2]
...
转发主节点[3]:PbftMsg [isOk = true, type = 0, node = 2, vnum = 0, time = 1551339457177, data = testD2, no = 0], onode = 2]
收到消息[1]:PbftMsg [isOk = true, type = 0, node = 3, vnum = 0, time = 1551339457177, data = testD3, no = 0], onode = 3]
...
转发主节点[1]:PbftMsg [isOk = true, type = 0, node = 3, vnum = 0, time = 1551339457177, data = testD3, no = 0], onode = 3]
转发主节点[2]:PbftMsg [isOk = true, type = 0, node = 3, vnum = 0, time = 1551339457177, data = testD3, no = 0], onode = 3]
```

收到消息[3]:PbftMsg [isOk = true, type = 0, node = 3, vnum = 0, time = 1551339457177, data = testD3, no = 0], onode = 3]
恢复[0]--------------
收到消息[0]:PbftMsg [isOk = true, type = 0, node = 1, vnum = 0, time = 1551339458178, data = testB1, no = 0], onode = 1]
…
请求执行成功[1]:PbftMsg [isOk = true, type = 3, node = 2, vnum = 0, time = 1551339458178, data = testB1, no = 2], onode = 1]
收到消息[2]:PbftMsg [isOk = true, type = 3, node = 2, vnum = 0, time = 1551339458178, data = testB1, no = 2], onode = 1]
请求执行成功[2]:PbftMsg [isOk = true, type = 4, node = 2, vnum = 0, time = 1551339458178, data = testB1, no = 2], onode = 1]
收到消息[0]:PbftMsg [isOk = true, type = 1, node = 0, vnum = 0, time = 1551339458178, data = testB1, no = 2], onode = 1]
….
请求执行成功[0]:PbftMsg [isOk = true, type = 4, node = 0, vnum = 0, time = 1551339458178, data = testB1, no = 2], onode = 1]
收到消息[0]:PbftMsg [isOk = true, type = 3, node = 0, vnum = 0, time = 1551339458178, data = testB1, no = 2], onode = 1]
请求主节点超时[3]:testD2
请求主节点超时[2]:testD3
收到消息[1]:PbftMsg [isOk = true, type = -2, node = 2, vnum = 1, time = 1551339458419, data = null, no = 0], onode = 2]
….
请求主节点超时[1]:testD3
收到消息[1]:PbftMsg [isOk = true, type = -2, node = 1, vnum = 1, time = 1551339458420, data = null, no = 0], onode = 1]
视图变更完成[1]: 1
收到消息[2]:PbftMsg [isOk = true, type = -2, node = 1, vnum = 1, time = 1551339458420, data = null, no = 0], onode = 1]
视图变更完成[2]: 1
请求重传[2]: PbftMsg [isOk = true, type = 0, node = 2, vnum = 1, time = 1551339457177, data = testD2, no = 0], onode = 2]
收到消息[1]:PbftMsg [isOk = true, type = 0, node = 2, vnum = 1, time = 1551339457177, data = testD2, no = 0], onode = 2]
…
异常消息[3]:PbftMsg [isOk = true, type = 2, node = 1, vnum = 1, time = 1551339457177, data = testD2, no = 3], onode = 2]
收到消息[2]:PbftMsg [isOk = true, type = 2, node = 2, vnum = 1, time = 1551339457177, data = testD2, no = 3], onode = 2]
请求主节点超时[1]:testD2
收到消息[3]:PbftMsg [isOk = true, type = 2, node = 2, vnum = 1, time = 1551339457177, data = testD2, no = 3], onode = 2]
异常消息[3]:PbftMsg [isOk = true, type = 2, node = 2, vnum = 1, time = 1551339457177, data = testD2, no = 3], onode = 2]
收到消息[3]:PbftMsg [isOk = true, type = -2, node = 1, vnum = 1, time = 1551339458420, data = null, no = 0], onode = 1]
…
视图变更完成[3]: 1

请求重传[3]:PbftMsg [isOk = true, type = 0, node = 3, vnum = 1, time = 1551339457177, data = testD3, no = 0], onode = 3]
收到消息[3]:PbftMsg [isOk = true, type = −2, node = 1, vnum = 2, time = 1551339458422, data = null, no = 0], onode = 1]
...
请求执行成功[1]:PbftMsg [isOk = true, type = 4, node = 1, vnum = 1, time = 1551339457177, data = testD3, no = 4], onode = 3]
收到消息[2]:PbftMsg [isOk = true, type = 3, node = 3, vnum = 1, time = 1551339457177, data = testD3, no = 4], onode = 3]
请求执行成功[2]:PbftMsg [isOk = true, type = 4, node = 2, vnum = 1, time = 1551339457177, data = testD3, no = 4], onode = 3]
收到消息[3]:PbftMsg [isOk = true, type = 3, node = 3, vnum = 1, time = 1551339457177, data = testD3, no = 4], onode = 3]
请求执行成功[3]:PbftMsg [isOk = true, type = 3, node = 3, vnum = 1, time = 1551339457177, data = testD3, no = 4], onode = 3]
收到消息[0]:PbftMsg [isOk = true, type = −2, node = 1, vnum = 1, time = 1551339458420, data = null, no = 0], onode = 1]
...
异常消息[0]:PbftMsg [isOk = true, type = 2, node = 1, vnum = 1, time = 1551339457177, data = testD2, no = 3], onode = 2]
收到消息[0]:PbftMsg [isOk = true, type = −2, node = 1, vnum = 2, time = 1551339458422, data = null, no = 0], onode = 1]
...
异常消息[0]:PbftMsg [isOk = true, type = 2, node = 1, vnum = 1, time = 1551339457177, data = testD3, no = 4], onode = 3]
收到消息[0]:PbftMsg [isOk = true, type = 3, node = 1, vnum = 1, time = 1551339457177, data = testD3, no = 4], onode = 3]
...
异常消息[0]:PbftMsg [isOk = true, type = 2, node = 3, vnum = 1, time = 1551339457177, data = testD3, no = 4], onode = 3]
收到消息[0]:PbftMsg [isOk = true, type = −2, node = 2, vnum = 1, time = 1551339458419, data = null, no = 0], onode = 2]
视图变更完成[0]: 1
收到消息[0]:PbftMsg [isOk = true, type = 3, node = 3, vnum = 1, time = 1551339457177, data = testD3, no = 4], onode = 3]
...
投票无效[2]:testD2|@|3
投票无效[1]:testD2|@|3

7.3 Raft 算法的实现

视频讲解

　　Raft 算法的解决方案可以理解成先在所有人中选出一个将军,再由将军来做决策进而形成共识。例如,有 A、B、C 三人,每人都有一个随机时间的倒计时器,倒计时一结束,这个人就会把自己当成将军候选人,然后派信使去问其他人,能否选自己为将军。假设现在 A 倒计时结束了,他派信使传递选举投票信息给 B 和 C,如果 B(或 C)还没把自己(B)当成候选人(即倒计时还没有结束)且没有把选票投给其他人(如 C),他就把票投给 A,信使回到 A 时,A 知道自己收到了足够票数成为了将军。每个人相当于一个分布式网络节点,每个节点有 Follower、

Candidate、Leader 三种状态,状态之间可以互相转换。

7.3.1 创建项目并添加依赖

用 IDEA 创建完项目 exraft 之后,确保在文件 pom.xml 中 < dependencies > 和 </dependencies > 之间添加了 Web、Jgroups 依赖,代码如例 7-19 所示。

【例 7-19】 添加 Web、Jgroups 依赖的代码示例。

```xml
<dependency>
            <groupId>org.springframework.boot</groupId>
            <artifactId>spring-boot-starter-web</artifactId>
</dependency>
<dependency>
            <groupId>jgroups</groupId>
            <artifactId>jgroups</artifactId>
            <version>2.9.0.GA</version>
</dependency>
```

7.3.2 创建类 Follower

在包 com.bookcode 中创建 algorithm 子包,并在包 com.bookcode.algorithm 中创建类 Follower,代码如例 7-20 所示。

【例 7-20】 类 Follower 的代码示例。

```java
package com.bookcode.algorithm;
import java.util.Timer;
import java.util.TimerTask;
import org.springframework.beans.factory.annotation.Value;
public class Follower {
    private Timer heartBeatMonitortimer = new Timer(true);
    private boolean heartbeatRcvd = false;
    private String name = "Follower ";
    public void setName(String name) {
        this.name = name;
    }
    @Value("${heartbeatTimeout}")
    public int beatTimeout;
    private void processVoteResp(ClusterMsg cl_msg) {
        System.out.println("Error.");
        return;
    }
    private void processVoteReq(ClusterMsg cl_msg) {
        int term = Integer.valueOf(cl_msg.getData());
        if (term > LeContext.getInstance().getTerm()) {
            LeContext.getInstance().setTerm(term);
```

```java
            MsgUtil.sendVoteRespMsg(cl_msg.getSrc());
            LeContext.getInstance().changeState("Candidate");
        } else {
            System.out.println("Error.");
        }
    }
    private void processHeartBeat(ClusterMsg cl_msg) {
        int term = Integer.valueOf(cl_msg.getData());
        if (term >= LeContext.getInstance().getTerm()) {
            heartbeatRcvd = true;
            if (term > LeContext.getInstance().getTerm())
                LeContext.getInstance().setTerm(term);
        } else {
            System.out.println("Similar to not getting heart beat");
        }
    }
    private void startHeartBeatMonitorThread() {
        int heartbeatTimeout = beatTimeout;
        heartBeatMonitortimer.scheduleAtFixedRate(new TimerTask() {
            @Override
            public void run() {
                if (heartbeatRcvd) {
                    heartbeatRcvd = false;
                    System.out.println("Heart Beat Received Properly..");
                } else {
                    System.out.println("No Heart Beat..Triggering election");
                    LeContext.getInstance().setTriggeredElection(true);
                    LeContext.getInstance().changeState("Candidate");
                }
            }
        }, heartbeatTimeout * 1000, heartbeatTimeout * 1000);
        System.out.println("Heartbeat Monitor Thread started");
    }
    private void stopHeartBeatMonitorThread() {
        heartBeatMonitortimer.cancel();
        heartBeatMonitortimer = null;
    }
    public String getName() {
        return this.name;
    }
    public String toString() {
        return "Follower: " + getName();
    }
}
```

7.3.3 创建类 Candidate

在包 com.bookcode.algorithm 中创建类 Candidate,代码如例 7-21 所示。

【例 7-21】 类 Candidate 的代码示例。

```java
package com.bookcode.algorithm;
import java.util.Random;
import java.util.Set;
import java.util.Timer;
import java.util.TimerTask;
import org.jgroups.*;
import org.springframework.beans.factory.annotation.Value;
public class Candidate {
    private Timer electionTimer;
    private boolean electionInProgress = false;
    private int currentElectionCnt = 0;
    private boolean triggeredElection = false;
    private Set<Address> respondedlist;
    private String name = "Candidate";
    public void setName(String name) {
        this.name = name;
    }
    @Value("${maxelectionlimit}")
    public int maxelectionlimit;
    @Value("${clusterSize}")
    public int clusterSize;
    @Value("${minelectionlimit}")
    public int minelectionlimit;
    private void processVoteResp(ClusterMsg cl_msg) {
        int term = Integer.valueOf(cl_msg.getData());
        if (!triggeredElection)
            return;
        System.out.println("process resp msg term = " + term + "Current term = " + LeContext.getInstance().getTerm());
        if (term == LeContext.getInstance().getTerm()) {
            System.out.println("adding to list..");
            respondedlist.add(cl_msg.getSrc());
            int respondedCnt = respondedlist.size();
            int ClusterSize = clusterSize;
            System.out.println("adding to list. respondedCnt = " + respondedCnt + "ClusterSize = " + ClusterSize);
            if ((respondedCnt + 1) * 2 > ClusterSize) {
                System.out.println("Reached success point...");
            }
        }
    }
    private void processVoteReq(ClusterMsg cl_msg) {
        int term = Integer.valueOf(cl_msg.getData());
        System.out.println("Request term = " + term);
        if (term > LeContext.getInstance().getTerm()) {
            LeContext.getInstance().setTerm(term);
            MsgUtil.sendVoteRespMsg(cl_msg.getSrc());
            LeContext.getInstance().setTriggeredElection(false);
```

```java
            } else {
                if(!triggeredElection && term == LeContext.getInstance().getTerm()) {
                    MsgUtil.sendVoteRespMsg(cl_msg.getSrc());
                }
            }
        }
        private void processHeartBeat(ClusterMsg cl_msg) {
            int term = Integer.valueOf(cl_msg.getData());
            if (term > LeContext.getInstance().getTerm()) {
                LeContext.getInstance().setTerm(term);
            } else if (term == LeContext.getInstance().getTerm()
                    && !triggeredElection) {
                LeContext.getInstance().changeState("Follower");
            }
        }
        private void startElection() {
            System.out.println("Election started..");
            currentElectionCnt++;
            if (currentElectionCnt > maxelectionlimit) {
                System.out.println("Error.");
            }
            electionInProgress = true;
            MsgUtil.sendVoteReqMsg();
            electionTimer = new Timer(true);
            int electionTimeOut = getElcetionTimeout();
            electionTimer.schedule(new TimerTask() {
                @Override
                public void run() {
                    System.out.println("Election Timeout task Entered");
                    if (electionInProgress) {
                        restartElection();
                    }
                }
            }, electionTimeOut * 1000);
        }
        private int getElcetionTimeout() {
            Random rand = new Random();
            int temp = rand.nextInt(maxelectionlimit - minelectionlimit + 1);
            return temp + minelectionlimit;
        }
        public void stopElection() {
            System.out.println("Election stopped");
            electionInProgress = false;
            electionTimer.cancel();
        }
        private void restartElection() {
            stopElection();
            startElection();
        }
        public String getName() {
```

```
            return this.name;
        }
        public String toString() {
            return "Candidate: " + getName();
        }
}
```

7.3.4 创建类 Leader

在包 com.bookcode.algorithm 中创建类 Leader,代码如例 7-22 所示。

【例 7-22】 类 Leader 的代码示例。

```
package com.bookcode.algorithm;
import java.util.Timer;
import java.util.TimerTask;
import org.springframework.beans.factory.annotation.Value;
public class Leader {
    private Timer heartBeatSendingtimer = new Timer(true);
    public void setName(String name) {
        this.name = name;
    }
    private String name = "Leader";
    @Value("${heartbeatFrequency}")
    public int beatFrequency;
    private void processVoteResp(ClusterMsg cl_msg) {
        System.out.println("Error.");
        return;
    }
    private void processVoteReq(ClusterMsg cl_msg) {
        int term = Integer.valueOf(cl_msg.getData());
        if (term > LeContext.getInstance().getTerm()) {
            MsgUtil.sendVoteRespMsg(cl_msg.getSrc());
            LeContext.getInstance().changeState("Candidate");
        } else if (term <= LeContext.getInstance().getTerm()) {
            MsgUtil.sendHearBeatMsg(cl_msg.getSrc());
        }
    }
    private void processHeartBeat(ClusterMsg cl_msg) {
        int term = Integer.valueOf(cl_msg.getData());
        if (term > LeContext.getInstance().getTerm()) {
            LeContext.getInstance().changeState("Follower");
        } else {
            MsgUtil.sendHearBeatMsg(cl_msg.getSrc());
        }
        return;
    }
    private void startHeartBeatSendingThread() {
```

```
            int heartbeatFrequency = beatFrequency;
            heartBeatSendingtimer.scheduleAtFixedRate(new TimerTask() {
                @Override
                public void run() {
                    System.out.println("Sending Heartbeat");
                    MsgUtil.sendHearBeatMsg();
                }
            }, 0, heartbeatFrequency * 1000);
            System.out.println("Heartbeat Sending Thread started");
    }
    private void stopHeartBeatSendingThread() {
        heartBeatSendingtimer.cancel();
    }
    public String getName() {
        return this.name;
    }
    public String toString() {
        return "Leader: " + getName();
    }
}
```

7.3.5 创建类 ClusterMsg

在包 com.bookcode.algorithm 中创建类 ClusterMsg,代码如例 7-23 所示。

【例 7-23】 类 ClusterMsg 的代码示例。

```
package com.bookcode.algorithm;
import java.io.Serializable;
import org.jgroups.Address;
import org.jgroups.Message;
public class ClusterMsg implements Serializable {
    public enum type {
        vote_req, vote_resp, heart_beat
    };
    private type msg_type;
    private String data;
    private Address src;
    public ClusterMsg(type msg_type, String data) {
        super();
        this.msg_type = msg_type;
        this.data = data;
        this.src = null;
    }
    public type getMsg_type() {
        return msg_type;
    }
    public void setMsg_type(type msg_type) {
```

```
        this.msg_type = msg_type;
    }
    public String getData() {
        return data;
    }
    public void setData(String data) {
        this.data = data;
    }
    public static Message getMsg(type ctype, String data) {
        Message msg = new Message();
        msg.setObject(new ClusterMsg(ctype, data));
        return msg;
    }
    public static ClusterMsg getClusterMsg(Message msg) {
        ClusterMsg cls_msg = (ClusterMsg) msg.getObject();
        cls_msg.setSrc(msg.getSrc());
        return cls_msg;
    }
    public void setSrc(Address src) {
        this.src = src;
    }
    public Address getSrc() {
        return this.src;
    }
}
```

7.3.6 创建类 MsgUtil

在包 com.bookcode.algorithm 中创建类 MsgUtil,代码如例 7-24 所示。

【例 7-24】 类 MsgUtil 的代码示例。

```
package com.bookcode.algorithm;
import org.jgroups.Address;
import org.jgroups.Message;
public class MsgUtil {
    public static void sendHearBeatMsg(Address dest) {
        System.out.println("send heart beat");
        Message tmp_msg = ClusterMsg.getMsg(ClusterMsg.type.heart_beat,
                String.valueOf(LeContext.getInstance().getTerm()));
        tmp_msg.setDest(dest);
    }
    public static void sendHearBeatMsg() {
        sendHearBeatMsg(null);
    }
    public static void sendVoteReqMsg() {
        System.out.println("send VoteReq Msg");
        Message tmp_msg = ClusterMsg.getMsg(ClusterMsg.type.vote_req,
```

```
                    String.valueOf(LeContext.getInstance().getTerm()));
        }
        public static void sendVoteRespMsg(Address dest) {
            System.out.println("send VoteResp Msg");
            Message tmp_msg = ClusterMsg.getMsg(ClusterMsg.type.vote_resp,
                    String.valueOf(LeContext.getInstance().getTerm()));
            tmp_msg.setDest(dest);
        }
}
```

7.3.7 创建类 RaftController

在包 com.bookcode 中创建 controller 子包，并在包 com.bookcode.controller 中创建类 RaftController，修改后的代码如例 7-25 所示。

【例 7-25】 类 RaftController 的代码示例。

```
package com.bookcode.controller;
import com.bookcode.algorithm.*;
import org.springframework.web.bind.annotation.GetMapping;
import org.springframework.web.bind.annotation.RestController;
@RestController
public class RaftController {
    @GetMapping("/testRaft")
    public String blockchain() {
        LeContext leContext = new LeContext("zs",4,true);
        leContext.changeState("lisi");
        System.out.println(leContext.toString());
        System.out.println(leContext.getTerm());
        MsgUtil.sendVoteReqMsg();
        Follower follower = new Follower();
        follower.setName("ww");
        System.out.println(follower.toString());
        Candidate candidate = new Candidate();
        candidate.setName("dd");
        System.out.println(candidate.toString());
        Leader leader = new Leader();
        leader.setName("ee");
        System.out.println(leader.toString());
        return "模拟 Raft 算法.";
    }
}
```

7.3.8 修改配置文件 application.properties

修改配置文件 application.properties，代码如例 7-26 所示。

【例 7-26】 修改后的配置文件 application.properties 代码示例。

```
heartbeatFrequency = 6
heartbeatTimeout = 5
maxelectionlimit = 4
clusterSize = 3
minelectionlimit = 2
```

7.3.9 运行程序

运行程序后,在浏览器中输入 localhost:8080/testRaft,浏览器中显示的结果如图 7-7 所示,控制台中的输出结果如 7-8 所示。

图 7-7 在浏览器中输入 localhost:8080/testRaft 后浏览器中显示的结果

```
from=zs state to=lisi
com.bookcode.algorithm.LeContext@c9979a
4
send VoteReq Msg
Follower: ww
Candidate: dd
Leader: ee
```

图 7-8 在浏览器中输入 localhost:8080/testRaft 后控制台中的结果

7.4 基于 PoW 的区块链应用示例

视频讲解

本节介绍如何实现基于 PoW 的区块链应用(联盟链)示例。

7.4.1 创建项目并添加依赖

用 IDEA 创建完项目 expbcmd 之后,确保在 pom.xml 文件中 < dependencies > 和 </dependencies > 之间添加了 Web 等依赖,代码如例 7-27 所示。

【例 7-27】 添加 Web 等依赖的代码示例。

```xml
<dependency>
        <groupId>org.springframework.boot</groupId>
        <artifactId>spring-boot-starter-web</artifactId>
</dependency>
```

```xml
<dependency>
    <groupId>org.springframework.boot</groupId>
    <artifactId>spring-boot-starter-data-jpa</artifactId>
</dependency>
<dependency>
    <groupId>mysql</groupId>
    <artifactId>mysql-connector-java</artifactId>
</dependency>
<dependency>
    <groupId>org.projectlombok</groupId>
    <artifactId>lombok</artifactId>
</dependency>
<dependency>
    <groupId>org.apache.commons</groupId>
    <artifactId>commons-lang3</artifactId>
</dependency>
<dependency>
    <groupId>commons-codec</groupId>
    <artifactId>commons-codec</artifactId>
</dependency>
```

7.4.2 创建类 BaseEntity

在包 com.bookcode 中创建 entity 子包，并在包 com.bookcode.entity 中创建类 BaseEntity，代码如例 7-28 所示。

【例 7-28】 类 BaseEntity 的代码示例。

```java
package com.bookcode.entity;
import lombok.Data;
import javax.persistence.GeneratedValue;
import javax.persistence.GenerationType;
import javax.persistence.Id;
import javax.persistence.MappedSuperclass;
import java.util.Date;
@MappedSuperclass
@Data
public class BaseEntity {
    @Id
    @GeneratedValue(strategy = GenerationType.IDENTITY)
    private Long id;
    private Date createTime;
    private Date updateTime;
}
```

7.4.3 创建类 Block

在包 com.bookcode.entity 中创建类 Block,代码如例 7-29 所示。

【例 7-29】 类 Block 的代码示例。

```java
package com.bookcode.entity;
import com.bookcode.utils.PowResult;
import com.bookcode.utils.ProofOfWork;
import lombok.AllArgsConstructor;
import lombok.Data;
import lombok.NoArgsConstructor;
import org.apache.commons.codec.binary.Hex;
import java.time.Instant;
@Data
@AllArgsConstructor
@NoArgsConstructor
public class Block {
    private static final String ZERO_HASH = Hex.encodeHexString(new byte[32]);
    private String hash;
    private String prevBlockHash;
    private String data;
    private long timeStamp;
    private long nonce;
    public static Block newGenesisBlock() {
        return Block.newBlock(ZERO_HASH, "Genesis Block");
    }
    public static Block newBlock(String previousHash, String data) {
        Block block = new Block("", previousHash, data, Instant.now().getEpochSecond(), 0);
        ProofOfWork pow = ProofOfWork.newProofOfWork(block);
        PowResult powResult = pow.run();
        block.setHash(powResult.getHash());
        block.setNonce(powResult.getNonce());
        return block;
    }
}
```

7.4.4 创建类 Blockchain

在包 com.bookcode.entity 中创建类 Blockchain,代码如例 7-30 所示。

【例 7-30】 类 Blockchain 的代码示例。

```java
package com.bookcode.entity;
import lombok.Data;
import javax.persistence.*;
import java.util.ArrayList;
```

```java
import java.util.List;
@Entity
@Table
@Data
public class Blockchain {
    @Id
    @GeneratedValue(strategy = GenerationType.IDENTITY)
    private Long id;
    private String groupId;
    private String blockchainsInfo;
    public Blockchain(String groupId, String blockchainsInfo) {
        this.groupId = groupId;
        this.blockchainsInfo = initNewBlock().toString();
    }
    public Blockchain() { }
    public Blockchain(String groupId) {
        this.groupId = groupId;
        this.blockchainsInfo = initNewBlock().toString();
    }
    private List<Block> initNewBlock(){
        List<Block> newblockchain = new ArrayList<Block>();
        newblockchain.add(Block.newGenesisBlock());
        return newblockchain;
    }
}
```

7.4.5 创建类 Member

在包 com.bookcode.entity 中创建类 Member,代码如例 7-31 所示。

【例 7-31】 类 Member 的代码示例。

```java
package com.bookcode.entity;
import lombok.Data;
import javax.persistence.Entity;
import javax.persistence.Table;
@Entity
@Table(name = "member")
@Data
public class Member extends BaseEntity {
    private String appId;
    private String name;
    private String ip;
    private String groupId;
}
```

7.4.6 创建类 MemberGroup

在包 com.bookcode.entity 中创建类 MemberGroup,代码如例 7-32 所示。

【例 7-32】 类 MemberGroup 的代码示例。

```
package com.bookcode.entity;
import lombok.Data;
import javax.persistence.Entity;
import javax.persistence.Table;
//联盟链,多个节点组成一个group,一个group为一个联盟链
@Entity
@Table(name = "member_group")
@Data
public class MemberGroup extends BaseEntity {
    private String name;
    //设置一个业务 ID
    private String groupId;
}
```

7.4.7 创建接口 MemberRepository

在包 com.bookcode 中创建 dao 子包,并在包 com.bookcode.dao 中创建接口 MemberRepository,代码如例 7-33 所示。

【例 7-33】 接口 MemberRepository 的代码示例。

```
package com.bookcode.dao;
import com.bookcode.entity.Member;
import org.springframework.data.jpa.repository.JpaRepository;
public interface MemberRepository extends JpaRepository<Member, Long> {
    Member findFirstByName(String name);
}
```

7.4.8 创建接口 MemberGroupRepository

在包 com.bookcode.dao 中创建接口 MemberGroupRepository,代码如例 7-34 所示。

【例 7-34】 接口 MemberGroupRepository 的代码示例。

```
package com.bookcode.dao;
import com.bookcode.entity.MemberGroup;
import org.springframework.data.jpa.repository.JpaRepository;
public interface MemberGroupRepository extends JpaRepository<MemberGroup, Long> {
}
```

7.4.9 创建接口 BlockchainRepository

在包 com.bookcode.dao 中创建接口 BlockchainRepository,代码如例 7-35 所示。

【例 7-35】 接口 BlockchainRepository 的代码示例。

```
package com.bookcode.dao;
import com.bookcode.entity.Blockchain;
import org.springframework.data.jpa.repository.JpaRepository;
public interface BlockchainRepository extends JpaRepository<Blockchain, Long> {
}
```

7.4.10 创建类 MemberService

在包 com.bookcode 中创建 service 子包,并在包 com.bookcode.service 中创建类 MemberService,代码如例 7-36 所示。

【例 7-36】 类 MemberService 的代码示例。

```
package com.bookcode.service;
import com.bookcode.entity.Member;
import com.bookcode.dao.MemberRepository;
import org.springframework.stereotype.Component;
import javax.annotation.Resource;
import java.util.List;
@Component
public class MemberService {
    @Resource
    private MemberRepository memberRepository;
    public String findGroupId(String memberName) {
        Member member = memberRepository.findFirstByName(memberName);
        if (member != null) {
            return member.getGroupId();
        }
        return null;
    }
    public List<Member> memberData() {
        List<Member> members = memberRepository.findAll();
        return members;
    }
}
```

7.4.11 创建类 MemberGroupService

在包 com.bookcode.service 中创建类 MemberGroupService,代码如例 7-37 所示。

【例7-37】 类MemberGroupService的代码示例。

```java
package com.bookcode.service;
import com.bookcode.entity.MemberGroup;
import com.bookcode.dao.MemberGroupRepository;
import org.springframework.stereotype.Component;
import javax.annotation.Resource;
import java.util.List;
@Component
public class MemberGroupService {
    @Resource
    private MemberGroupRepository memberGroupRepository;
    public List<MemberGroup> memberGroupData() {
        List<MemberGroup> memberGroups = memberGroupRepository.findAll();
        return memberGroups;
    }
}
```

7.4.12　创建类BlockchainService

在包com.bookcode.service中创建类BlockchainService，代码如例7-38所示。

【例7-38】 类BlockchainService的代码示例。

```java
package com.bookcode.service;
import com.bookcode.entity.Blockchain;
import com.bookcode.dao.BlockchainRepository;
import org.springframework.stereotype.Component;
import javax.annotation.Resource;
import java.util.List;
@Component
public class BlockchainService {
    @Resource
    private BlockchainRepository blockchainRepository;
    public String findBlockchainInfoByGroupId(String groupId){
        Blockchain bc = new Blockchain("2");
        blockchainRepository.save(bc);
        List<Blockchain> bLists = blockchainRepository.findAll();
        String blockchainInfos = "区块链信息如下：\n";
        if(!bLists.isEmpty()){
            for(int i = 0;i < bLists.size();i++){
                if(groupId.equals(bLists.get(i).getGroupId())){
                    blockchainInfos += bLists.get(i).getBlockchainsInfo();
                }
            }
        }
        else{
            blockchainInfos += "无区块链信息.";
```

```
            }
            return blockchainInfos;
    }
}
```

7.4.13 创建类 ByteUtils

在包 com.bookcode 中创建 utils 子包,并在包 com.bookcode.utils 中创建类 ByteUtils,代码如例 7-39 所示。

【例 7-39】 类 ByteUtils 的代码示例。

```
package com.bookcode.utils;
import org.apache.commons.lang3.ArrayUtils;
import java.nio.ByteBuffer;
import java.util.Arrays;
import java.util.stream.Stream;
public class ByteUtils {
    //将多字节数组合并成一字节数组
    public static byte[] merge(byte[]... bytes) {
        Stream<Byte> stream = Stream.of();
        for (byte[] b : bytes) {
            stream = Stream.concat(stream, Arrays.stream(ArrayUtils.toObject(b)));
        }
        return ArrayUtils.toPrimitive(stream.toArray(Byte[]::new));
    }
    //long 类型转换为 byte[]
    public static byte[] toBytes(long val) {
        return ByteBuffer.allocate(Long.BYTES).putLong(val).array();
    }
}
```

7.4.14 创建类 ProofOfWork

在包 com.bookcode.utils 中创建类 ProofOfWork,代码如例 7-40 所示。

【例 7-40】 类 ProofOfWork 的代码示例。

```
package com.bookcode.utils;
import com.bookcode.entity.Block;
import lombok.Data;
import org.apache.commons.codec.digest.DigestUtils;
import org.apache.commons.lang3.StringUtils;
import java.math.BigInteger;
@Data
public class ProofOfWork {
```

```java
    //难度目标位
    public static final int TARGET_BITS = 26;
    private Block block;
    //难度目标值
    private BigInteger target;
    private ProofOfWork(Block block, BigInteger target) {
        this.block = block;
        this.target = target;
    }
    //对1进行移位运算,将1向左移动(256 - TARGET_BITS)位,得到难度目标值
    public static ProofOfWork newProofOfWork(Block block) {
        BigInteger targetValue = BigInteger.ONE.shiftLeft((256 - TARGET_BITS));
        return new ProofOfWork(block, targetValue);
    }
    //运行工作量证明,开始挖矿,找到小于难度目标值的Hash
    public PowResult run() {
        long nonce = 0;
        String shaHex = "";
        System.out.printf("Mining the block containing: %s \n", this.getBlock().getData());
        long startTime = System.currentTimeMillis();
        while (nonce < Long.MAX_VALUE) {
            byte[] data = this.prepareData(nonce);
            shaHex = DigestUtils.sha256Hex(data);
            if (new BigInteger(shaHex, 16).compareTo(this.target) == -1) {
System.out.printf("Elapsed Time: %s seconds \n", (float) (System.currentTimeMillis() - startTime) / 1000);
                System.out.printf("correct hash Hex: %s \n\n", shaHex);
                break;
            } else {
                nonce++;
            }
        }
        return new PowResult(nonce, shaHex);
    }
    public boolean validate() {
        byte[] data = this.prepareData(this.getBlock().getNonce());
        return new BigInteger(DigestUtils.sha256Hex(data), 16).compareTo(this.target) == -1;
    }
    //在准备区块数据时,从原始数据类型转换为byte[],不直接从字符串进行转换
    private byte[] prepareData(long nonce) {
        byte[] prevBlockHashBytes = {};
        if (StringUtils.isNoneBlank(this.getBlock().getPrevBlockHash())) {
prevBlockHashBytes = new BigInteger(this.getBlock().getPrevBlockHash(), 16).toByteArray();
        }
        return ByteUtils.merge(
                prevBlockHashBytes,
                this.getBlock().getData().getBytes(),
                ByteUtils.toBytes(this.getBlock().getTimeStamp()),
```

```
                ByteUtils.toBytes(TARGET_BITS),
                ByteUtils.toBytes(nonce)
        );
    }
}
```

7.4.15　创建类 PowResult

在包 com.bookcode.utils 中创建类 PowResult,代码如例 7-41 所示。

【例 7-41】　类 PowResult 的代码示例。

```
package com.bookcode.utils;
import lombok.AllArgsConstructor;
import lombok.Data;
//工作量计算结果
@Data
@AllArgsConstructor
public class PowResult {
    //计数器
    private long nonce;
    private String hash;
}
```

7.4.16　创建类 MemberandGroupController

在包 com.bookcode 中创建 controller 子包,并在包 com.bookcode.controller 中创建类 MemberandGroupController,代码如例 7-42 所示。

【例 7-42】　类 MemberandGroupController 的代码示例。

```
package com.bookcode.controller;
import com.bookcode.service.BlockchainService;
import com.bookcode.service.MemberGroupService;
import com.bookcode.service.MemberService;
import org.springframework.web.bind.annotation.GetMapping;
import org.springframework.web.bind.annotation.PathVariable;
import org.springframework.web.bind.annotation.RestController;
import javax.annotation.Resource;
@RestController
public class MemberandGroupController {
    @Resource
    private MemberService memberService;
    @Resource
    private MemberGroupService memberGroupService;
    @Resource
```

```
    private BlockchainService blockchainService;
    @GetMapping("/members")
    public String allmembers(){
        return memberService.memberData().toString();
    }
    @GetMapping("/blockchains/{groupid}")
    public String allblockchainsofgroup(@PathVariable String groupid){
        return blockchainService.findBlockchainInfoByGroupId(groupid);
    }
    @GetMapping("/groups")
    public String allgroups(){
        return memberGroupService.memberGroupData().toString();
    }
}
```

7.4.17 创建配置文件 application.yml

在目录 src/main/resources 下创建配置文件 application.yml 并修改配置文件代码，修改后的代码如例 7-43 所示。

【例 7-43】 修改后的配置文件 application.yml 代码示例。

```
spring:
  jpa:
    database: mysql
    show-sql: true
    hibernate:
      ddl-auto: update
      naming:
        physical-strategy: org.springframework.boot.orm.jpa.hibernate.SpringPhysicalNamingStrategy
    database-platform: org.hibernate.dialect.MySQL5InnoDBDialect
    #不加这句则默认为 myisam 引擎
  profiles:
    active: ${ENV:local}
  datasource:
    driver-class-name: com.mysql.cj.jdbc.Driver
    url: jdbc:mysql://localhost:3306/testnew?serverTimezone=GMT%2B8&useUnicode=true&characterEncoding=UTF-8&useSSL=false
    password: sa
    username: root
logging:
  file: ./logback.log
```

7.4.18 运行程序

运行程序后，在浏览器中输入 localhost:8080/members，浏览器的输出结果如图 7-9 所示。在浏览器中输入 localhost:8080/groups,浏览器的输出结果如图 7-10 所示。在浏览器中输入 localhost:8080/blockchains/2,浏览器的输出结果如图 7-11 所示。

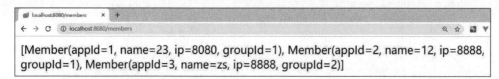

图 7-9　在浏览器中输入 localhost:8080/members 后浏览器的输出结果

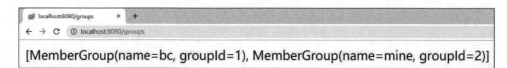

图 7-10　在浏览器中输入 localhost:8080/groups 后浏览器的输出结果

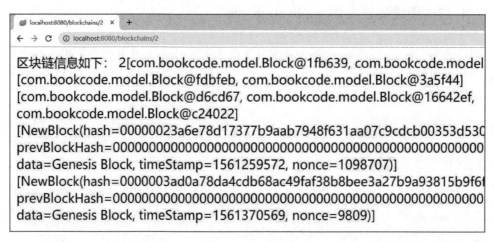

图 7-11　在浏览器中输入 localhost:8080/blockchains/2 后浏览器的输出结果

习题 7

实验题

1. 实现 PoW 算法。
2. 实现 PBFT 算法。
3. 实现 Raft 算法。

第8章 工作流、多语言和智能合约的应用

本章主要介绍工作流 Activiti 的应用、Spring Cloud Sidecar 多语言的应用智能合约的模拟实现和基于 Activiti 的区块链应用、示例。

8.1 Activiti 的应用

视频讲解

Activiti 业务流程管理(BPM)开源项目于 2010 年启动,旨在支持新的 BPMN 2.0 标准,为互操作性、云架构等提供技术实现。本节介绍工作流 Activiti 的应用。

8.1.1 创建项目并添加依赖

用 IDEA 创建完项目 activitiexample 之后,确保在文件 pom.xml 中<dependencies>和</dependencies>之间添加了 Web、H2 数据库依赖,代码如例 8-1 所示。

【例 8-1】 添加 Web、H2 数据库依赖的代码示例。

```
<dependency>
    <groupId>org.springframework.boot</groupId>
    <artifactId>spring-boot-starter-web</artifactId>
</dependency>
<dependency>
    <groupId>com.h2database</groupId>
    <artifactId>h2</artifactId>
</dependency>
<dependency>
    <groupId>org.activiti</groupId>
    <artifactId>activiti-spring-boot-starter-basic</artifactId>
    <version>5.19.0.2</version>
</dependency>
```

```
</dependency>
<dependency>
        <groupId>org.springframework.boot</groupId>
        <artifactId>spring-boot-starter-data-jpa</artifactId>
</dependency>
```

8.1.2 创建类 Applicant

在包 com.bookcode 中创建 entity 子包,并在包 com.bookcode.entity 中创建类 Applicant,代码如例 8-2 所示。

【例 8-2】 类 Applicant 的代码示例。

```
package com.bookcode.entity;
import javax.persistence.Entity;
import javax.persistence.GeneratedValue;
import javax.persistence.Id;
@Entity
public class Applicant {
    @Id
    @GeneratedValue
    private Long id;
    private String name;
    private String email;
    private String phoneNumber;
    public Applicant() {
    }
    public Applicant(String name, String email, String phoneNumber) {
        this.name = name;
        this.email = email;
        this.phoneNumber = phoneNumber;
    }
    public Long getId() {
        return id;
    }
    public void setId(Long id) {
        this.id = id;
    }
    public String getName() {
        return name;
    }
    public void setName(String name) {
        this.name = name;
    }
    public String getEmail() {
        return email;
    }
    public void setEmail(String email) {
```

```
        this.email = email;
    }
    public String getPhoneNumber() {
        return phoneNumber;
    }
    public void setPhoneNumber(String phoneNumber) {
        this.phoneNumber = phoneNumber;
    }
}
```

8.1.3 创建接口 ApplicantRepository

在包 com.bookcode 中创建 repository 子包,并在包 com.bookcode.repository 中创建接口 ApplicantRepository,代码如例 8-3 所示。

【例 8-3】 接口 ApplicantRepository 的代码示例。

```
package com.bookcode.repository;
import com.bookcode.entity.Applicant;
import org.springframework.data.jpa.repository.JpaRepository;
public interface ApplicantRepository extends JpaRepository<Applicant, Long> {
}
```

8.1.4 创建类 ResumeService

在包 com.bookcode 中创建 service 子包,并在包 com.bookcode.service 中创建类 ResumeService,代码如例 8-4 所示。

【例 8-4】 类 ResumeService 的代码示例。

```
package com.bookcode.service;
import org.springframework.stereotype.Component;
@Component
public class ResumeService {
    public void storeResume() {
        System.out.println("Storing resume ...");
    }
}
```

8.1.5 创建类 HireProcessRestController

在包 com.bookcode 中创建 controller 子包,并在包 com.bookcode.controller 中创建类 HireProcessRestController,代码如例 8-5 所示。

【例 8-5】 类 HireProcessRestController 的代码示例。

```
package com.bookcode.controller;
import com.bookcode.entity.Applicant;
import com.bookcode.repository.ApplicantRepository;
import org.activiti.engine.RuntimeService;
import org.springframework.beans.factory.annotation.Autowired;
import org.springframework.http.HttpStatus;
import org.springframework.http.MediaType;
import org.springframework.web.bind.annotation.*;
import java.util.Collections;
import java.util.Map;
@RestController
public class HireProcessRestController {
    @Autowired
    private RuntimeService runtimeService;
    @Autowired
    private ApplicantRepository applicantRepository;
    @ResponseStatus(value = HttpStatus.OK)
    @RequestMapping(value = "/start-hire-process", method = RequestMethod.POST,
            produces = MediaType.APPLICATION_JSON_VALUE)
    public void startHireProcess(@RequestBody Map<String, String> data) {
        Applicant applicant = new Applicant(data.get("name"), data.get("email"), data.get("phoneNumber"));
        applicantRepository.save(applicant);
        Map<String, Object> vars = Collections.<String, Object>singletonMap("applicant", applicant);
        runtimeService.startProcessInstanceByKey("hireProcessWithJpa", vars);
    }
}
```

8.1.6 修改配置文件 application.properties

修改配置文件 application.properties，修改后的代码如例 8-6 所示。

【例 8-6】 修改后的配置文件 application.properties 代码示例。

```
spring.h2.console.enabled = true
```

8.1.7 修改入口类

修改入口类，修改后的入口类代码如例 8-7 所示。

【例 8-7】 修改后的入口类代码示例。

```
import org.springframework.boot.autoconfigure.SpringBootApplication;
import org.springframework.context.annotation.Bean;
@SpringBootApplication(exclude = SecurityAutoConfiguration.class)
```

```
public class ActivitiexampleApplication {
    @Bean
    InitializingBean usersAndGroupsInitializer(final IdentityService identityService) {
        return new InitializingBean() {
            public void afterPropertiesSet() throws Exception {
                Group group = identityService.newGroup("user");
                group.setName("users");
                group.setType("security-role");
                identityService.saveGroup(group);
                User admin = identityService.newUser("admin");
                admin.setPassword("admin");
                identityService.saveUser(admin);
            }
        };
    }
    public static void main(String[] args) {
        SpringApplication.run(ActivitiexampleApplication.class, args);
    }
}
```

8.1.8 修改测试类

修改测试类,修改后的代码如例 8-8 所示。

【例 8-8】 修改后的测试类代码示例。

```
package com.bookcode;
import com.bookcode.repository.ApplicantRepository;
import com.bookcode.entity.Applicant;
import org.activiti.engine.HistoryService;
import org.activiti.engine.RuntimeService;
import org.activiti.engine.TaskService;
import org.activiti.engine.runtime.ProcessInstance;
import org.activiti.engine.task.Task;
import org.junit.After;
import org.junit.Assert;
import org.junit.Before;
import org.junit.Test;
import org.junit.runner.RunWith;
import org.springframework.beans.factory.annotation.Autowired;
import org.springframework.boot.test.context.SpringBootTest;
import org.springframework.test.context.junit4.SpringRunner;
import java.util.HashMap;
import java.util.List;
import java.util.Map;
import org.subethamail.wiser.Wiser;
@RunWith(SpringRunner.class)
@SpringBootTest
```

```java
public class ActivitiexampleApplicationTests {
    @Autowired
    private RuntimeService runtimeService;
    @Autowired
    private TaskService taskService;
    @Autowired
    private HistoryService historyService;
    @Autowired
    private ApplicantRepository applicantRepository;
     private Wiser wiser;
    @Before
    public void setup() {
        wiser = new Wiser();
      wiser.setPort(1025);
        wiser.start();
    }
    @After
    public void cleanup() {
        wiser.stop();
    }
    @Test
    public void contextLoads() {
    }
    @Test
    public void testHappyPath() {
        Applicant applicant = new Applicant("John Doe", "john@activiti.org", "12344");
        applicantRepository.save(applicant);
        Map<String, Object> variables = new HashMap<String, Object>();
        variables.put("applicant", applicant);
        ProcessInstance processInstance = runtimeService.startProcessInstanceByKey
("hireProcessWithJpa", variables);
        Task task = taskService.createTaskQuery()
                .processInstanceId(processInstance.getId())
                .taskCandidateGroup("dev-managers")
                .singleResult();
        Assert.assertEquals("Telephone interview", task.getName());
        Map<String, Object> taskVariables = new HashMap<String, Object>();
        taskVariables.put("telephoneInterviewOutcome", true);
        taskService.complete(task.getId(), taskVariables);
        List<Task> tasks = taskService.createTaskQuery()
                .processInstanceId(processInstance.getId())
                .orderByTaskName().asc()
                .list();
        Assert.assertEquals(2, tasks.size());
        Assert.assertEquals("Financial negotiation", tasks.get(0).getName());
        Assert.assertEquals("Tech interview", tasks.get(1).getName());
        taskVariables = new HashMap<String, Object>();
        taskVariables.put("techOk", true);
        taskService.complete(tasks.get(0).getId(), taskVariables);
        taskVariables = new HashMap<String, Object>();
        taskVariables.put("financialOk", true);
        taskService.complete(tasks.get(1).getId(), taskVariables);
        Assert.assertEquals(1,wiser.getMessages().size());
```

```
                Assert.assertEquals(1, historyService.createHistoricProcessInstanceQuery().
finished().count());
    }
}
```

8.1.9 运行程序

从网上下载并向项目添加 subethasmtp-smtp-1.2.jar 和 subethasmtp-wiser-1.2.jar 文件，添加后的结果如图 8-1 所示。运行入口类后打开 Postman 工具，再在 Postman 工具 URL 处输入 http://127.0.0.1:8080/start-hire-process，选择 POST 方法，单击 Send 按钮，如图 8-2 所示。这样就会在数据库对应的表中增加一条记录，并且在控制台中输出一行 "Storing resume..."，如图 8-3 所示。重复执行两次相同操作就可以在数据库对应的表中增加两条记录。为了更好地观察对数据库的访问情况，可以通过在浏览器中输入 localhost:8080/h2-console 访问 H2 数据库的控制台，结果如图 8-4 所示。运行测试类，控制台的主要输出如图 8-5 所示。

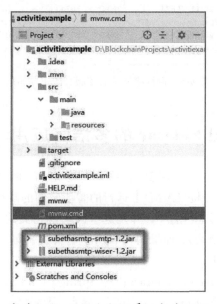

图 8-1　向项目添加 subethasmtp-smtp-1.2.jar 和 subethasmtp-wiser-1.2.jar 的结果

图 8-2　在 Postman 中增加一条记录

```
Storing resume ...
```

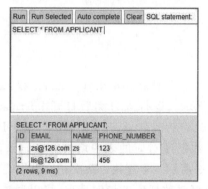

图 8-3　在 Postman 中增加一条记录后的控制台输出

图 8-4　在 Postman 中增加两条记录后 H2 数据库的信息

```
Storing resume ...
2019-02-28    20:11:53.624         INFO     10740     ---     [rver.SMTPServer]
org.subethamail.smtp.server.SMTPServer      : SMTP Server socket shut down.
2019-02-28    20:11:53.640         INFO     10740     ---     [rver.SMTPServer]
org.subethamail.smtp.server.SMTPServer      : SMTP Server socket shut down.
```

图 8-5　运行测试类后控制台的主要输出

8.2　Spring Cloud Sidecar 的多语言应用

视频讲解

区块链技术中用到许多不同的语言，例如，Solidity 开发语言是在 Ethereum 虚拟机(EVM)上运行的智能契约的高级语言，Hyperledger 用到了 Go 语言。假如要用 Java 开发这些区块链相关的应用时，需要用到转换机制(即 Sidecar 模式)；如果用 Web3j 进行开发。第 9 章内容可以视为是 Sidecar 模式的具体化。

可使用 Spring Cloud Netflix Sidecar(简称 Spring Cloud Sidecar)很方便地整合非 JVM 服务。非 JVM 服务可操作 Eureka 的 REST 端点，从而实现服务注册与发现。Spring Cloud Netflix Sidecar 的灵感来自 Netflix Prana，它包括了一个简单的 HTTP API 来获取指定服务所有实例信息(例如主机和端口)。

8.2.1　创建项目并添加依赖

用 IDEA 创建完项目 sidecarexample 之后，文件 pom. xml 的代码如例 8-9 所示。

【例 8-9】　文件 pom. xml 的代码示例。

```
<?xml version = "1.0" encoding = "UTF - 8"?>
```

```xml
<project xmlns="http://maven.apache.org/POM/4.0.0" xmlns:xsi="http://www.w3.org/2001/XMLSchema-instance"
    xsi:schemaLocation="http://maven.apache.org/POM/4.0.0 http://maven.apache.org/xsd/maven-4.0.0.xsd">
    <modelVersion>4.0.0</modelVersion>
    <parent>
        <groupId>org.springframework.boot</groupId>
        <artifactId>spring-boot-starter-parent</artifactId>
        <version>2.1.2.RELEASE</version>
        <relativePath/> <!-- lookup parent from repository -->
    </parent>
    <groupId>com.bookcode</groupId>
    <artifactId>demo</artifactId>
    <version>0.0.1-SNAPSHOT</version>
    <name>demo</name>
    <description>Demo project for Spring Cloud</description>
    <properties>
        <java.version>1.8</java.version>
        <spring-cloud.version>Greenwich.RELEASE</spring-cloud.version>
    </properties>
    <dependencies>
        <dependency>
            <groupId>org.springframework.cloud</groupId>
            <artifactId>spring-cloud-netflix-sidecar</artifactId>
        </dependency>
        <dependency>
            <groupId>org.springframework.cloud</groupId>
            <artifactId>spring-cloud-starter-netflix-eureka-client</artifactId>
        </dependency>
        <dependency>
            <groupId>org.springframework.boot</groupId>
            <artifactId>spring-boot-starter-test</artifactId>
            <scope>test</scope>
        </dependency>
    </dependencies>
    <dependencyManagement>
        <dependencies>
            <dependency>
                <groupId>org.springframework.cloud</groupId>
                <artifactId>spring-cloud-dependencies</artifactId>
                <version>${spring-cloud.version}</version>
                <type>pom</type>
                <scope>import</scope>
            </dependency>
        </dependencies>
    </dependencyManagement>
    <build>
        <plugins>
            <plugin>
                <groupId>org.springframework.boot</groupId>
```

```xml
            <artifactId>spring-boot-maven-plugin</artifactId>
        </plugin>
    </plugins>
</build>
<repositories>
    <repository>
        <id>spring-milestones</id>
        <name>Spring Milestones</name>
        <url>https://repo.spring.io/milestone</url>
    </repository>
</repositories>
</project>
```

8.2.2 修改配置文件 application.properties

修改配置文件 application.properties，修改后的代码如例 8-10 所示。

【例 8-10】 修改后的配置文件 application.properties 代码示例。

```
server.port = 8082
spring.application.name = sidecar-server
eureka.client.service-url.defaultZone = http://localhost:8761/eureka/
sidecar.port = 8205
sidecar.health-uri = http://localhost:${sidecar.port}/health.json
management.endpoints.web.exposure.include = *
```

8.2.3 修改入口类

修改入口类，修改后的入口类代码如例 8-11 所示。

【例 8-11】 修改后的入口类代码示例。

```java
package com.bookcode;
import org.springframework.boot.SpringApplication;
import org.springframework.boot.autoconfigure.SpringBootApplication;
import org.springframework.cloud.netflix.sidecar.EnableSidecar;
@EnableSidecar
@SpringBootApplication
public class DemoApplication {
    public static void main(String[] args) {
        SpringApplication.run(DemoApplication.class, args);
    }
}
```

8.2.4 创建文件 node-service.js

创建文件 node-service.js 并修改文件代码,修改后的代码如例 8-12 所示。

【例 8-12】 修改后的文件 node-service.js 的代码示例。

```javascript
//nodejs 引入 http、url、path 模块
var http = require('http');
var url = require("url");
var path = require('path');
//创建 server
var server = http.createServer(function(req, res) {
//获得请求的路径
  var pathname = url.parse(req.url).pathname;
  res.writeHead(200, { 'Content-Type' : 'application/json; charset=utf-8' });
  if (pathname === '/') {
    res.end(JSON.stringify({ "index" : "欢迎来到简单异构系统之 nodejs 服务首页" }));
  }
  else if (pathname === '/health.json') {
    res.end(JSON.stringify({ "status" : "UP" }));
  }
  else {
    res.end("404");
  }
});
//创建监听,并打印日志
server.listen(8205, function() {
  console.log('开始监听本地端口:8205');
});
```

8.2.5 Spring Cloud Eureka 注册中心的实现

用 IDEA 创建完项目 eureka-server 之后,文件 pom.xml 的代码如例 8-13 所示。

【例 8-13】 文件 pom.xml 的代码示例。

```xml
<?xml version="1.0" encoding="UTF-8"?>
<project xmlns="http://maven.apache.org/POM/4.0.0" xmlns:xsi="http://www.w3.org/2001/XMLSchema-instance"
         xsi:schemaLocation="http://maven.apache.org/POM/4.0.0 http://maven.apache.org/xsd/maven-4.0.0.xsd">
    <modelVersion>4.0.0</modelVersion>
    <groupId>com.example</groupId>
    <artifactId>demo</artifactId>
    <version>0.0.1-SNAPSHOT</version>
    <packaging>jar</packaging>
    <name>demo</name>
```

```xml
<description>Demo project for Spring Boot</description>
<parent>
    <groupId>org.springframework.boot</groupId>
    <artifactId>spring-boot-starter-parent</artifactId>
    <version>2.1.2.RELEASE</version>
    <relativePath/> <!-- lookup parent from repository -->
</parent>
<properties>
    <project.build.sourceEncoding>UTF-8</project.build.sourceEncoding>
    <project.reporting.outputEncoding>UTF-8</project.reporting.outputEncoding>
    <java.version>1.8</java.version>
    <spring-cloud.version>Greenwich.RELEASE</spring-cloud.version>
</properties>
<dependencies>
    <dependency>
        <groupId>org.springframework.cloud</groupId>
        <artifactId>spring-cloud-starter-netflix-eureka-server</artifactId>
    </dependency>
    <dependency>
        <groupId>org.springframework.boot</groupId>
        <artifactId>spring-boot-starter-test</artifactId>
        <scope>test</scope>
    </dependency>
</dependencies>
<dependencyManagement>
    <dependencies>
        <dependency>
            <groupId>org.springframework.cloud</groupId>
            <artifactId>spring-cloud-dependencies</artifactId>
            <version>${spring-cloud.version}</version>
            <type>pom</type>
            <scope>import</scope>
        </dependency>
    </dependencies>
</dependencyManagement>
<build>
    <plugins>
        <plugin>
            <groupId>org.springframework.boot</groupId>
            <artifactId>spring-boot-maven-plugin</artifactId>
        </plugin>
    </plugins>
</build>
<repositories>
    <repository>
        <id>sonatype-nexus-snapshots</id>
        <name>Sonatype Nexus Snapshots</name>
        <url>http://maven.aliyun.com/nexus/content/groups/public</url>
```

```xml
            <releases>
                <enabled>false</enabled>
            </releases>
            <snapshots>
                <enabled>true</enabled>
            </snapshots>
        </repository>
        <repository>
            <id>spring-milestones</id>
            <name>Spring Milestones</name>
            <url>https://repo.spring.io/milestone</url>
            <snapshots>
                <enabled>false</enabled>
            </snapshots>
        </repository>
    </repositories>
</project>
```

在目录 src/main/resources 下，创建配置文件 application.yml 并修改配置文件代码，修改后的代码如例 8-14 所示。

【例 8-14】 修改后的配置文件 application.yml 代码示例。

```yaml
server:
  port: 8761    #指定该 Eureka 实例的端口
eureka:
  instance:
    hostname: localhost
  client:
    registerWithEureka: false    #表示不注册到 Eureka 注册中心，因为本应用是 Eureka 注册中心
    fetchRegistry: false
    serviceUrl:
      defaultZone: http://${eureka.instance.hostname}:${server.port}/eureka/
```

修改入口类，修改后的入口类代码如例 8-15 所示。

【例 8-15】 修改后的入口类代码示例。

```java
package com.bookcode;
import org.springframework.boot.SpringApplication;
import org.springframework.boot.autoconfigure.SpringBootApplication;
import org.springframework.cloud.netflix.eureka.server.EnableEurekaServer;
@SpringBootApplication
@EnableEurekaServer
public class DemoApplication {
    public static void main(String[] args) {
        SpringApplication.run(DemoApplication.class, args);
    }
}
```

8.2.6 运行程序

执行启动命令,代码如例 8-16 所示(请根据情况修改为 node-service.js 文件所在目录),启动程序 node-service.js。

【例 8-16】 启动程序 node-service.js 的命令示例。

```
node D:\BlockchainProjects\ch8-2\sidecarexample\src\main\resources\node-service.js
```

依次运行 eureka-server 程序和 sidercarexample 程序。在浏览器中输入 localhost:8205,结果如图 8-6 所示。在浏览器中输入 localhost:8205/health.json,结果如图 8-7 所示。在浏览器中输入 localhost:8082,结果如图 8-8 所示。在浏览器中输入 localhost:8082/ping,结果如图 8-9 所示。在浏览器中输入 localhost:8082/hosts/sidecar-server,结果如图 8-10 所示。在浏览器中输入 localhost:8082/actuator/health,结果如图 8-11 所示。关闭 node-service.js 程序,在浏览器中输入 localhost:8082/actuator/health,结果如图 8-12 所示。

图 8-6 在浏览器中输入 localhost:8205 的结果

图 8-7 在浏览器中输入 localhost:8205/health.json 的结果

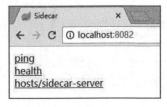

图 8-8 在浏览器中输入 localhost:8082 的结果

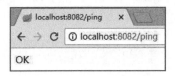

图 8-9 在浏览器中输入 localhost:8082/ping 的结果

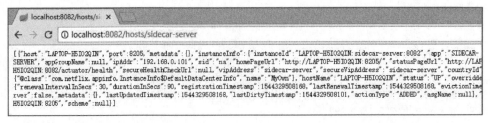

图 8-10 在浏览器中输入 localhost:8082/hosts/sidecar-server 的结果

第8章 工作流、多语言和智能合约的应用

图 8-11 在浏览器中输入 localhost：8082/actuator/health 的结果（没有关闭 node-service.js 程序）

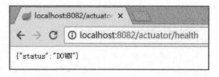

图 8-12 在浏览器中输入 localhost：8082/actuator/health 的结果（关闭 node-service.js 服务后）

8.3 智能合约的模拟实现

本节介绍结合具体应用模拟智能合约的实现。

视频讲解

8.3.1 创建项目并添加依赖

用 IDEA 创建完项目 exsc 之后，确保在文件 pom.xml 中 < dependencies > 和 </dependencies> 之间添加了 Web 等依赖，代码如例 8-17 所示。

【例 8-17】 添加 Web 等依赖的代码示例。

```
<dependency>
        <groupId>org.springframework.boot</groupId>
        <artifactId>spring-boot-starter-web</artifactId>
</dependency>
<dependency>
        <groupId>mysql</groupId>
        <artifactId>mysql-connector-java</artifactId>
</dependency>
<dependency>
        <groupId>org.springframework.boot</groupId>
        <artifactId>spring-boot-starter-data-jpa</artifactId>
</dependency>
<dependency>
        <groupId>org.projectlombok</groupId>
        <artifactId>lombok</artifactId>
</dependency>
```

8.3.2 创建类 AccountNew

在包 com.bookcode 中创建 entity 子包，并在包 com.bookcode.entity 中创建类 AccountNew，代码如例 8-18 所示。

【例 8-18】 类 AccountNew 的代码示例。

```
package com.bookcode.entity;
```

```
import lombok.Data;
import javax.persistence.*;
@Entity
@Table(name = "t_account")
@Data
public class AccountNew {
    @Id
    @GeneratedValue(strategy = GenerationType.AUTO)
    private Long id;
    private String balance;
    private String keyvalue;
    private String name;
}
```

8.3.3 创建接口 AccountRepository

在包 com.bookcode 中创建 repository 子包,并在包 com.bookcode.repository 中创建接口 AccountRepository,代码如例 8-19 所示。

【例 8-19】 接口 AccountRepository 的代码示例。

```
package com.bookcode.repository;
import com.bookcode.entity.AccountNew;
import org.springframework.data.jpa.repository.JpaRepository;
public interface AccountRepository extends JpaRepository<AccountNew, Long> {
}
```

8.3.4 创建类 SmartContractStub

在包 com.bookcode 中创建 smartcontract 子包,并在包 com.bookcode.smartcontract 中创建类 SmartContractStub,代码如例 8-20 所示。

【例 8-20】 类 SmartContractStub 的代码示例。

```
package com.bookcode.smartcontract;
public class SmartContractStub {
    public String getStringState(String key) {
        return key;
    }
    public void putState(String key, byte[] value) {
        if (key == null) {
            throw new NullPointerException("key cannot be null");
        }
        if (key.length() == 0) {
            throw new IllegalArgumentException("key cannot not be an empty string");
```

            }
        }
}
```

### 8.3.5 创建接口 ISmartContract

在包 com.bookcode.smartcontract 中创建接口 ISmartContract，代码如例 8-21 所示。

【例 8-21】 接口 ISmartContract 的代码示例。

```java
package com.bookcode.smartcontract;
import java.util.HashMap;
import java.util.Map;
public interface ISmartContract {
    class SmartContractResponse {
        private final Status status;
        private final String message;
        private final byte[] payload;
        public SmartContractResponse(Status status, String message, byte[] payload) {
            this.status = status;
            this.message = message;
            this.payload = payload;
        }
        public String getMessage() {
            return message;
        }
        public enum Status {
            SUCCESS(200),
            INTERNAL_SERVER_ERROR(500),
            ERRORTHRESHOLD(600);
            private static final Map<Integer, Status> codeToStatus = new HashMap<>();
            static {
                for (Status status : Status.values()) {
                    codeToStatus.put(status.code, status);
                }
            }
            private final int code;
            Status(int code) {
                this.code = code;
            }
        }
    }
}
```

8.3.6 创建类 SCController

在包 com.bookcode 中创建 controller 子包，并在包 com.bookcode.controller 中创建
```

类 SCController,代码如例 8-22 所示。

【例 8-22】 类 SCController 的代码示例。

```java
package com.bookcode.controller;
import com.bookcode.entity.AccountNew;
import com.bookcode.repository.AccountRepository;
import com.bookcode.smartcontract.ISmartContract;
import com.bookcode.smartcontract.SmartContractStub;
import org.springframework.beans.factory.annotation.Autowired;
import org.springframework.web.bind.annotation.GetMapping;
import org.springframework.web.bind.annotation.PathVariable;
import org.springframework.web.bind.annotation.RestController;
import java.math.BigDecimal;
import java.nio.charset.StandardCharsets;
import java.util.List;
import static java.lang.String.format;
@RestController
public class SCController {
    @Autowired
    private AccountRepository accountRepository;
    protected static ISmartContract.SmartContractResponse newSuccessResponse(String message, byte[] payload) {
        return new ISmartContract.SmartContractResponse(ISmartContract.SmartContractResponse.Status.SUCCESS, message, payload);
    }
    protected static ISmartContract.SmartContractResponse newSuccessResponse(String message) {
        return newSuccessResponse(message, null);
    }
    protected static ISmartContract.SmartContractResponse newErrorResponse(String message) {
        return newErrorResponse(message, null);
    }
    protected static ISmartContract.SmartContractResponse newErrorResponse(String message, byte[] payload) {
        return new ISmartContract.SmartContractResponse(
                ISmartContract.SmartContractResponse.Status.INTERNAL_SERVER_ERROR, message, payload);
    }
    @GetMapping("/testSCTransfer/{from}/{to}/{value}")
    public String move(SmartContractStub stub, @PathVariable String from, @PathVariable String to, @PathVariable String value) {
        String fromKey = from;
        String toKey = to;
        String amount = value;
        System.out.println("\n================================");
        System.out.println("转账人:" + fromKey + " 收款人:" + toKey + " 金额:" + amount);
        //获取身份信息
        final String fromKeyState = stub.getStringState(fromKey);
```

```java
            final String toKeyState = stub.getStringState(toKey);
            System.out.println("\n================================= ");
            System.out.println("从账户" + fromKeyState + "转到账户" + toKeyState + ".");
            if (fromKey.equals(toKey)) {
                return newErrorResponse("Please do not transfer money to yourself.").getMessage();
            }
            //转账人余额获取类型转换
            BigDecimal fromAccountBalance = new BigDecimal(getBalance(fromKeyState));
            BigDecimal toAccountBalance = new BigDecimal(getBalance(toKeyState));
            //转账金额类型转换
            BigDecimal transferAmount = new BigDecimal(Integer.parseInt(amount));
            //确保金额足够
            if (transferAmount.compareTo(fromAccountBalance) > 0) {
                return newErrorResponse("Lack of funds").getMessage();
            }
            if (transferAmount.compareTo(BigDecimal.valueOf(0)) < 0) {
                return newErrorResponse("The transfer amount must be greater than zero").getMessage();
            } else {
                //转账操作
                System.out.println("\n=================================" + format
("%s transfer %s to %s", fromKey, transferAmount.toString(), toKey));
                BigDecimal newFromAccountBalance = fromAccountBalance.subtract(transferAmount);
                BigDecimal newToAccountBalance = toAccountBalance.add(transferAmount);
                updateBalance(newFromAccountBalance.toString(), fromKey);
                updateBalance(newToAccountBalance.toString(), toKey);
                System.out.println("\n=================================" + format
("balance: %s = %s, %s = %s", fromKey, newFromAccountBalance.toString(), toKey,
newToAccountBalance.toString()));
                stub.putState(fromKey, newFromAccountBalance.toString().getBytes
(StandardCharsets.UTF_8));
                stub.putState(toKey, newToAccountBalance.toString().getBytes(StandardCharsets.
UTF_8));
                return newSuccessResponse("成功转账" + transferAmount.toString() + "元.").
getMessage();
            }
        }
    }
    void updateBalance(String newBalance, String fromKeyState) {
        List<AccountNew> accountNewList = accountRepository.findAll();
        for (AccountNew accountNew : accountNewList) {
            if (accountNew.getKeyvalue().equals(fromKeyState)) {
                AccountNew accountNew1 = new AccountNew();
                accountNew1.setId(accountNew.getId());
                accountNew1.setKeyvalue(accountNew.getKeyvalue());
                accountNew1.setName(accountNew.getName());
                accountNew1.setBalance(newBalance);
                accountRepository.save(accountNew1);
            }
        }
```

```java
    }
    @GetMapping("/testSCQuery/{account}")
    public String query(SmartContractStub stub, @PathVariable String account) {
        final String fromKeyState = stub.getStringState(account);
        final String fromKeyName = getStringName(account);
        BigDecimal fromAccountBalance = new BigDecimal(getBalance(fromKeyState));
        String message = "账户余额信息:[{\"name\":\"" + fromKeyName + "\",\"value\":"
+ fromAccountBalance.toString() + "}]";
        return newSuccessResponse(message).getMessage();
    }
    private Integer getBalance(String fromKeyState) {
        List < AccountNew > accountNewList = accountRepository.findAll();
        for (AccountNew accountNew : accountNewList) {
            if (accountNew.getKeyvalue().equals(fromKeyState))
                return Integer.parseInt(accountNew.getBalance());
        }
        return 20000;
    }
    public String getStringName(String key) {
        List < AccountNew > accountNewList = accountRepository.findAll();
        for (AccountNew accountNew : accountNewList) {
            if (accountNew.getKeyvalue().equals(key))
                return accountNew.getName();
        }
        return "某人";
    }
}
```

8.3.7 修改配置文件 application.properties

修改配置文件 application.properties，修改后的代码如例 8-23 所示。

【例 8-23】 修改后的配置文件 application.properties 代码示例。

```
spring.datasource.driver-class-name = com.mysql.cj.jdbc.Driver
spring.datasource.username = root
spring.datasource.password = sa
spring.jpa.hibernate.ddl-auto = update
spring.datasource.url = jdbc:mysql://localhost:3306/testnew?serverTimezone = GMT%
2B8&useUnicode = true&characterEncoding = UTF-8&useSSL = false
spring.jpa.database-platform:org.hibernate.dialect.MySQL5InnoDBDialect
```

8.3.8 运行程序

运行程序后，在浏览器中输入 localhost:8080/testSCQuery/001，结果如图 8-13 所示。在浏览器中输入 localhost:8080/testSCTransfer/001/002/1000，结果如图 8-14 所示；该操

作使得数据库中的数据由如图 8-15 所示变化为如图 8-16 所示。

图 8-13 在浏览器中输入 localhost：8080/testSCQuery/001 的结果

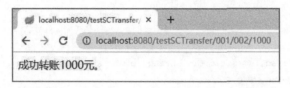

图 8-14 在浏览器中输入 localhost：8080/testSCTransfer/001/002/1000 的结果

图 8-15 在浏览器中输入 localhost：8080/testSCTransfer/001/002/1000 之前数据库中的数据

图 8-16 在浏览器中输入 localhost：8080/testSCTransfer/001/002/1000 之后数据库中的数据

8.4 基于 Activiti 的区块链应用示例

本节介绍如何实现基于 Activiti 的区块链应用(审批)示例。

视频讲解

8.4.1 创建项目并添加依赖

用 IDEA 创建完项目 exabc 之后，确保在 pom.xml 文件中 < dependencies > 和 </dependencies >之间添加了 Web 等依赖，代码如例 8-24 所示。

【例 8-24】 添加 Web 等依赖的代码示例。

```
< dependency >
        < groupId > org.springframework.boot </groupId >
        < artifactId > spring - boot - starter - web </artifactId >
</dependency >
< dependency >
        < groupId > org.springframework.boot </groupId >
        < artifactId > spring - boot - starter - data - jpa </artifactId >
```

```xml
        </dependency>
        <dependency>
            <groupId>mysql</groupId>
            <artifactId>mysql-connector-java</artifactId>
        </dependency>
        <dependency>
            <groupId>org.projectlombok</groupId>
            <artifactId>lombok</artifactId>
        </dependency>
        <dependency>
            <groupId>org.activiti</groupId>
            <artifactId>activiti-spring-boot-starter-basic</artifactId>
            <version>6.0.0</version>
        </dependency>
        <dependency>
            <groupId>org.springframework.boot</groupId>
            <artifactId>spring-boot-starter-security</artifactId>
        </dependency>
```

8.4.2 创建类 Person

在包 com.bookcode 中创建 entity 子包,并在包 com.bookcode.entity 中创建类 Person,代码如例 8-25 所示。

【例 8-25】 类 Person 的代码示例。

```java
package com.bookcode.entity;
import lombok.*;
import javax.persistence.*;
import java.io.Serializable;
@NoArgsConstructor
@Data
@Entity
public class Person implements Serializable {
    @Id
    @GeneratedValue
    private Long personId;
    private String personName;
    @ManyToOne(targetEntity = Comp.class)
    private Comp comp;
    public Person(String personName) {
        this.personName = personName;
    }
}
```

8.4.3 创建类 Comp

在包 com.bookcode.entity 中创建类 Comp,代码如例 8-26 所示。

【例 8-26】 类 Comp 的代码示例。

```java
package com.bookcode.entity;
import lombok.*;
import javax.persistence.*;
import java.io.Serializable;
import java.util.List;
@NoArgsConstructor
@Data
@Entity
public class Comp implements Serializable {
    @Id
    @GeneratedValue
    private Long compId;
    private String compName;
    @OneToMany(mappedBy = "comp", targetEntity = Person.class)
    private List<Person> people;
    public Comp(String compName)
    {this.compName = compName;}
}
```

8.4.4 创建类 Block

在包 com.bookcode.entity 中创建类 Block,代码如例 8-27 所示。

【例 8-27】 类 Block 的代码示例。

```java
package com.bookcode.entity;
import lombok.Data;
import java.util.Date;
@Data
public class Block {
    private int hash;
    private String data;
    private long timestamp;
    public Block(String data) {
        this.data = data;
        this.timestamp = new Date().getTime();
        this.hash = data.hashCode();
    }
}
```

8.4.5 创建类 Blockchain

在包 com.bookcode.entity 中创建类 Blockchain,代码如例 8-28 所示。

【例 8-28】 类 Blockchain 的代码示例。

```
package com.bookcode.entity;
import lombok.*;
import javax.persistence.*;
@AllArgsConstructor
@NoArgsConstructor
@Data
@Table(name = "t_bcInfo")
@Entity
public class Blockchain {
    @Id
    @GeneratedValue
    private Long id;
    private String personName;
    private String content;
    public Blockchain(String personName){
        this.personName = personName;
        Block block = new Block(personName);
        this.content = "\n" + block.toString();
    }
}
```

8.4.6 创建类 TaskRepresentation

在包 com.bookcode.entity 中创建类 TaskRepresentation,代码如例 8-29 所示。

【例 8-29】 类 TaskRepresentation 的代码示例。

```
package com.bookcode.entity;
import lombok.*;
@AllArgsConstructor
@Data
public class TaskRepresentation {
    private String id;
    private String name;
}
```

8.4.7 创建接口 PersonRepository

在包 com.bookcode 中创建 dao 子包,并在包 com.bookcode.dao 中创建接口 Person-

Repository,代码如例 8-30 所示。

【例 8-30】 接口 PersonRepository 的代码示例。

```
package com.bookcode.dao;
import com.bookcode.entity.Person;
import org.springframework.data.jpa.repository.JpaRepository;
public interface PersonRepository extends JpaRepository<Person, Long> {
    public Person findByPersonName(String personName);
}
```

8.4.8 创建接口 CompRepository

在包 com.bookcode.dao 中创建接口 CompRepository,代码如例 8-31 所示。

【例 8-31】 接口 CompRepository 的代码示例。

```
package com.bookcode.dao;
import com.bookcode.entity.Comp;
import org.springframework.data.jpa.repository.JpaRepository;
public interface CompRepository extends JpaRepository<Comp, Long> {
}
```

8.4.9 创建接口 BlockchainRepository

在包 com.bookcode.dao 中创建接口 BlockchainRepository,代码如例 8-32 所示。

【例 8-32】 接口 BlockchainRepository 的代码示例。

```
package com.bookcode.dao;
import com.bookcode.entity.Blockchain;
import org.springframework.data.jpa.repository.JpaRepository;
public interface BlockchainRepository extends JpaRepository<Blockchain, Long> {
}
```

8.4.10 创建类 ActiveService

在包 com.bookcode 中创建 service 子包,并在包 com.bookcode.service 中创建类 ActiveService,代码如例 8-33 所示。

【例 8-33】 类 ActiveService 的代码示例。

```
package com.bookcode.service;
import org.activiti.engine.*;
import org.activiti.engine.task.Task;
```

```java
import org.springframework.beans.factory.annotation.Autowired;
import org.springframework.stereotype.Service;
import javax.transaction.Transactional;
import java.util.*;
@Service
@Transactional
public class ActivitiService {
@Autowired
private RuntimeService runtimeService;
@Autowired
private TaskService taskService;
//开始流程,传入申请者的 ID 以及公司的 ID
public void startProcess(Long personId, Long compId) {
    Map<String, Object> variables = new HashMap<String, Object>();
    variables.put("personId", personId);
    variables.put("compId", compId);
    runtimeService.startProcessInstanceByKey("joinProcess", variables);
}
//获得某个人的任务列表
public List<Task> getTasks(String assignee) {
    return taskService.createTaskQuery().taskCandidateUser(assignee).list();
}
//完成任务
public void completeTasks(Boolean joinApproved, String taskId, Long money) {
    Map<String, Object> taskVariables = new HashMap<String, Object>();
    taskVariables.put("joinApproved", joinApproved);
    taskVariables.put("money", money);
    taskService.complete(taskId, taskVariables);
}
}
```

8.4.11　创建类 JoinService

在包 com.bookcode.service 中创建类 JoinService,代码如例 8-34 所示。

【例 8-34】　类 JoinService 的代码示例。

```java
package com.bookcode.service;
import com.bookcode.entity.*;
import com.bookcode.dao.*;
import org.activiti.engine.delegate.DelegateExecution;
import org.springframework.beans.factory.annotation.Autowired;
import org.springframework.stereotype.Service;
import java.util.*;
@Service
public class JoinService {
@Autowired
    PersonRepository personRepository;
```

```java
@Autowired
private CompRepository compRepository;
//加入公司操作,可从DelegateExecution获取流程中的变量
public void joinGroup(DelegateExecution execution) {
    Boolean bool = (Boolean) execution.getVariable("joinApproved");
    if (bool) {
        Long personId = (Long) execution.getVariable("personId");
        Long compId = (Long) execution.getVariable("compId");
        Comp comp = findByCompId(compId);
        Person person = findByPersonId(personId);
        person.setComp(comp);
        personRepository.save(person);
        System.out.println("加入组织成功");
    } else {
        System.out.println("加入组织失败");
    }
}
private Person findByPersonId(Long personId) {
    List<Person> personList = personRepository.findAll();
    if(personList.isEmpty()){
        return null;
    }
    else {
        for(int i = 0;i < personList.size();i++){
            if(personId == personList.get(i).getPersonId()){
                Person person = new Person();
                person.setPersonId(personId);
                person.setPersonName(personList.get(i).getPersonName());
                person.setComp(personList.get(i).getComp());
                return person;
            }
            else{
                continue;
            }
        }
    }
    return null;
}
private Comp findByCompId(Long compId) {
    List<Comp> compList = compRepository.findAll();
    if(compList.isEmpty()){
        return null;
    }
    else {
        for(int i = 0;i < compList.size();i++){
            if(compId == compList.get(i).getCompId()){
                Comp comp = new Comp();
                comp.setCompId(compId);
                comp.setCompName(compList.get(i).getCompName());
                comp.setPeople(compList.get(i).getPeople());
```

```
                    return comp;
                }
                else{
                    continue;
                }
            }
        }
        return null;
    }
    public void nonJoinGroup(DelegateExecution execution) {
        System.out.println("金额异常,加入组织失败!");
    }
    //获取符合条件的审批人
    public List<String> findUsers(DelegateExecution execution) {
        return Arrays.asList("admin", "wtr");
    }
}
```

8.4.12　创建类 MyRestController

在包 com.bookcode 中创建 controller 子包,并在包 com.bookcode.controller 中创建类 MyRestController,代码如例 8-35 所示。

【例 8-35】　类 MyRestController 的代码示例。

```
package com.bookcode.controller;
import com.bookcode.entity.*;
import com.bookcode.dao.*;
import com.bookcode.service.ActivitiService;
import org.activiti.engine.task.Task;
import org.springframework.beans.factory.annotation.Autowired;
import org.springframework.web.bind.annotation.*;
import java.util.*;
@RestController
public class MyRestController {
@Autowired
private ActivitiService myService;
@Autowired
    BlockchainRepository blockchainRepository;
@Autowired
    PersonRepository personRepository;
//开启流程实例
@GetMapping("/processes/{personId}/{compId}")
public String startProcessInstance (@PathVariable Long personId, @PathVariable Long compId) {
        myService.startProcess(personId, compId);
        Person person = personRepository.getOne(personId);
        blockchainRepository.save(new Blockchain(person.getPersonName()));
```

```java
            return "开启流程示例成功";
    }
    @GetMapping("/blockchains")
        public String getChains(){
            List<Blockchain> ls = blockchainRepository.findAll();
            String sbc = "区块链信息为: ";
            sbc += ls.toString();
                return sbc;
        }
    //获取当前人的任务
    @GetMapping("/tasks")
    public List<TaskRepresentation> getTasks(@RequestParam String assignee) {
        List<Task> tasks = myService.getTasks(assignee);
        List<TaskRepresentation> dtos = new ArrayList<TaskRepresentation>();
        for (Task task : tasks) {
            dtos.add(new TaskRepresentation(task.getId(), task.getName()));
        }
        return dtos;
    }
    //完成任务
    @GetMapping("/complete/{joinApproved}/{taskId}/{money}")
     public String complete (@PathVariable Boolean joinApproved, @PathVariable String taskId, @PathVariable Long money) {
        myService.completeTasks(joinApproved, taskId, money);
        return "完成任务.";
    }
}
```

8.4.13　创建文件 join.bpmn20.xml

在目录 src/main/resources/static 下创建子目录 processes，在目录 src/main/resources/static/processes 下创建文件 join.bpmn20.xml 并修改文件代码，修改后的代码如例 8-36 所示。

【例 8-36】　修改后的文件 join.bpmn20.xml 的代码示例。

```xml
<?xml version = "1.0" encoding = "UTF-8" standalone = "yes"?>
<definitions xmlns = "http://www.omg.org/spec/BPMN/20100524/MODEL" xmlns:activiti = "http://activiti.org/bpmn" xmlns:bpmndi = "http://www.omg.org/spec/BPMN/20100524/DI" xmlns:dc = "http://www.omg.org/spec/DD/20100524/DC" xmlns:di = "http://www.omg.org/spec/DD/20100524/DI" xmlns:tns = "http://www.activiti.org/test" xmlns:xsd = "http://www.w3.org/2001/XMLSchema" xmlns:xsi = "http://www.w3.org/2001/XMLSchema-instance" expressionLanguage = "http://www.w3.org/1999/XPath" id = "m1557205519781" name = "" targetNamespace = "http://www.activiti.org/test" typeLanguage = "http://www.w3.org/2001/XMLSchema">
    <process id = "joinProcess" isClosed = "false" isExecutable = "true" name = "Join process" processType = "None">
        <startEvent id = "startevent1" name = "Start">
            <extensionElements>
```

```xml
        <activiti:formProperty id="personId" name="person id" required="true" type="long"/>
        <activiti:formProperty id="compId" name="company Id" required="true" type="long"/>
      </extensionElements>
    </startEvent>
    <endEvent id="endevent1" name="End"/>
    <userTask activiti:candidateUsers="${joinService.findUsers(execution)}" activiti:exclusive="true" id="ApprovalTask" isForCompensation="true" name="Approval Task">
      <extensionElements>
        <activiti:formProperty id="joinApproved" name="Join Approved" type="enum">
          <activiti:value id="true" name="Approve"/>
          <activiti:value id="false" name="Reject"/>
        </activiti:formProperty>
      </extensionElements>
    </userTask>
    <sequenceFlow id="flow1" sourceRef="startevent1" targetRef="ApprovalTask"/>
    <serviceTask activiti:exclusive="true" activiti:expression="${joinService.joinGroup(execution)}" id="AutoTask" name="Auto Task"/>
    <sequenceFlow id="flow2" sourceRef="ApprovalTask" targetRef="AutoTask"/>
    <sequenceFlow id="flow3" sourceRef="AutoTask" targetRef="endevent1"/>
  </process>
  <bpmndi:BPMNDiagram documentation="background=#3C3F41;count=1;horizontalcount=1;orientation=0;width=842.4;height=1195.2;imageableWidth=832.4;imageableHeight=1185.2;imageableX=5.0;imageableY=5.0" id="Diagram-_1" name="New Diagram">
    <bpmndi:BPMNPlane bpmnElement="joinProcess">
      <bpmndi:BPMNShape bpmnElement="AutoTask" id="Shape-AutoTask">
        <dc:Bounds height="55.0" width="85.0" x="257.0" y="188.5"/>
        <bpmndi:BPMNLabel>
          <dc:Bounds height="55.0" width="85.0" x="0.0" y="0.0"/>
        </bpmndi:BPMNLabel>
      </bpmndi:BPMNShape>
      <bpmndi:BPMNShape bpmnElement="ApprovalTask" id="Shape-ApprovalTask">
        <dc:Bounds height="55.0" width="85.0" x="112.0" y="188.5"/>
        <bpmndi:BPMNLabel>
          <dc:Bounds height="55.0" width="85.0" x="0.0" y="0.0"/>
        </bpmndi:BPMNLabel>
      </bpmndi:BPMNShape>
      <bpmndi:BPMNShape bpmnElement="endevent1" id="Shape-endevent1">
        <dc:Bounds height="32.0" width="32.0" x="402.0" y="200.0"/>
        <bpmndi:BPMNLabel>
          <dc:Bounds height="32.0" width="32.0" x="0.0" y="0.0"/>
        </bpmndi:BPMNLabel>
      </bpmndi:BPMNShape>
      <bpmndi:BPMNShape bpmnElement="startevent1" id="Shape-startevent1">
        <dc:Bounds height="32.0" width="32.0" x="20.0" y="200.0"/>
        <bpmndi:BPMNLabel>
          <dc:Bounds height="32.0" width="32.0" x="0.0" y="0.0"/>
        </bpmndi:BPMNLabel>
      </bpmndi:BPMNShape>
```

```xml
        < bpmndi:BPMNEdge bpmnElement = "flow1" id = "BPMNEdge_flow1" sourceElement =
"startevent1" targetElement = "ApprovalTask">
          < di:waypoint x = "52.0" y = "216.0"/>
          < di:waypoint x = "112.0" y = "216.0"/>
          < bpmndi:BPMNLabel >
            < dc:Bounds height = " - 1.0" width = " - 1.0" x = " - 1.0" y = " - 1.0"/>
          </bpmndi:BPMNLabel >
        </bpmndi:BPMNEdge >
        < bpmndi:BPMNEdge bpmnElement = "flow2" id = "BPMNEdge_flow2" sourceElement =
"ApprovalTask" targetElement = "AutoTask">
          < di:waypoint x = "197.0" y = "216.0"/>
          < di:waypoint x = "257.0" y = "216.0"/>
          < bpmndi:BPMNLabel >
            < dc:Bounds height = " - 1.0" width = " - 1.0" x = " - 1.0" y = " - 1.0"/>
          </bpmndi:BPMNLabel >
        </bpmndi:BPMNEdge >
        < bpmndi:BPMNEdge bpmnElement = "flow3" id = "BPMNEdge_flow3" sourceElement = "AutoTask"
targetElement = "endevent1">
          < di:waypoint x = "342.0" y = "216.0"/>
          < di:waypoint x = "402.0" y = "216.0"/>
          < bpmndi:BPMNLabel >
            < dc:Bounds height = " - 1.0" width = " - 1.0" x = " - 1.0" y = " - 1.0"/>
          </bpmndi:BPMNLabel >
        </bpmndi:BPMNEdge >
      </bpmndi:BPMNPlane >
    </bpmndi:BPMNDiagram >
</definitions >
```

8.4.14 修改配置文件 application.properties

修改配置文件 application.properties，修改后的代码如例 8-37 所示。

【例 8-37】 修改后的配置文件 application.properties 代码示例。

```
spring.datasource.driver - class - name = com.mysql.cj.jdbc.Driver
spring.datasource.username = root
spring.datasource.password = sa
spring.jpa.hibernate.ddl - auto = update
spring.datasource.url = jdbc:mysql://localhost:3306/testnew? serverTimezone = GMT %
2B8&useUnicode = true&characterEncoding = UTF - 8&useSSL = false
spring.jpa.database - platform: org.hibernate.dialect.MySQL5InnoDBDialect
spring.security.user.name = zs
spring.security.user.password = zs
```

8.4.15 修改入口类

修改入口类，修改后的入口类代码如例 8-38 所示。

【例 8-38】 修改后的入口类代码示例。

```java
package com.bookcode;
import com.bookcode.entity.*;
import com.bookcode.dao.*;
import com.bookcode.service.ActivitiService;
import org.activiti.spring.boot.SecurityAutoConfiguration;
import org.springframework.beans.factory.annotation.Autowired;
import org.springframework.boot.CommandLineRunner;
import org.springframework.boot.SpringApplication;
import org.springframework.boot.autoconfigure.SpringBootApplication;
import org.springframework.boot.autoconfigure.domain.EntityScan;
import org.springframework.context.annotation.Bean;
import org.springframework.context.annotation.ComponentScan;
import org.springframework.data.jpa.repository.config.EnableJpaRepositories;
@EnableJpaRepositories(basePackages = {"com.bookcode.dao"})
@ComponentScan(value = {"com.bookcode.*","com.bookcode.dao"})
@EntityScan("com.bookcode.entity")
@SpringBootApplication(exclude = SecurityAutoConfiguration.class)
public class ExabcApplication {
    @Autowired
    private PersonRepository personRepository;
    @Autowired
    private CompRepository compRepository;
    //初始化模拟数据
    @Bean
    public CommandLineRunner init(final ActivitiService myService) {
        return new CommandLineRunner() {
            public void run(String... strings) throws Exception {
                if (personRepository.findAll().size() == 0) {
                    personRepository.save(new Person("wtr"));
                    personRepository.save(new Person("wyf"));
                    personRepository.save(new Person("admin"));
                }
                if (compRepository.findAll().size() == 0) {
                    Comp group = new Comp("great company");
                    compRepository.save(group);
                    Person admin = personRepository.findByPersonName("admin");
                    Person wtr = personRepository.findByPersonName("wtr");
                    admin.setComp(group);
                    wtr.setComp(group);
                    personRepository.save(admin);
                    personRepository.save(wtr);
                }
            }
        };
    }
    public static void main(String[] args) {
        SpringApplication.run(ExabcApplication.class, args);
    }
}
```

8.4.16 运行程序

运行程序后，在浏览器中输入 localhost:8080/processes/3/1，浏览器自动跳转到如图 8-17 所示的安全登录页面。输入正确的 Username 和 Password（本例中均为 zs，参考配置文件的设置）后单击 Sign in 按钮，自动跳转到如图 8-18 所示的页面。在浏览器中输入 localhost:8080/complete/true/7507/2，浏览器的输出结果如图 8-19 所示。在浏览器中输入 localhost:8080/blockchains，浏览器的输出结果如图 8-20 所示。

图 8-17　在浏览器中输入 localhost:8080/processes/3/1 之后跳转到的安全登录页面

图 8-18　输入正确 Username 和 Password 后单击 Sign in 按钮浏览器自动跳转到的页面

图 8-19　在浏览器中输入 localhost:8080/complete/true/7507/2 的结果

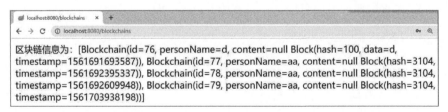

图 8-20　在浏览器中输入 localhost:8080/blockchains 的结果

习题 8

实验题

1. 实现 Activiti 的应用开发。
2. 实现 Spring Cloud Sidecar 的应用开发。
3. 实现智能合约的模拟实现。

第9章

区块链库、API和框架的应用

本章介绍 bitcoinj、fabric-sdk-java、eth-contract-api、exonum-java-binding、web3j、WavesJ 等应用和基于 web3j 钱包业务功能的示例。

9.1 bitcoinj 的应用

9.1.1 bitcoinj 简介

视频讲解

bitcoinj 库是 Bitcoin 协议的开源 Java 实现,可以用它开发与比特币网络交互的 Java 应用程序。bitcoinj 不是原始 Bitcoin 客户端的完整实现,而是更轻量级和可访问的版本。它由 core、crypto、jni、kits、net、params、protocols、script、signers、store、uri、utils、wallet 等包构成;每个包又包含了不同的类,如 kits 由 WalletAppKit、package-info 两个类构成。WalletAppKit 有序地指挥着各个模块的工作。

bitcoinj 的应用程序使用以下对象:

(1) NetworkParameters 实例,用于选择用户所在的网络(生产或测试)。

(2) 用于存储 ECKey 和其他数据的 Wallet 实例。

(3) 用于管理网络连接的 PeerGroup 实例。

(4) 一个 BlockChain 实例,它管理共享的全局数据结构,使 Bitcoin 工作。

(5) 一个 BlockStore 实例,它将块链数据结构保存在某个位置,就像在磁盘上一样。

(6) WalletEventListener 实现,用于接收钱包事件。

为了简化设置,还有一个 WalletAppKit 对象可以创建上述对象并将它们连接在一起。

更多信息可以参考 https://github.com/bitcoinj/bitcoinj 中的文档和具体代码。

9.1.2 创建项目并添加依赖

用 IDEA 创建完项目 jcoinapi 之后,确保在文件 pom.xml 中<dependencies>和</dependencies>之间添加了 bitcoinj-jopt 依赖,代码如例 9-1 所示。

【例 9-1】 添加 bitcoinj-jopt 依赖的代码示例。

```xml
<dependency>
        <groupId>org.bitcoinj</groupId>
        <artifactId>bitcoinj-core</artifactId>
        <version>0.14.7</version>
</dependency>
<dependency>
        <groupId>net.sf.jopt-simple</groupId>
        <artifactId>jopt-simple</artifactId>
        <version>5.0.3</version>
</dependency>
```

9.1.3 创建类 BitcoinJController

在包 com.bookcode 中创建 jcoinapi.controller 子包,并在包 com.bookcode.jcoinapi.controller 中创建类 BitcoinJController,代码如例 9-2 所示。注意,导入的类 LegacyAddress 原来在包 org.bitcoinj.core 中,为了测试的需要把它粘贴到了本项目的 com.bookcode.jcoinapi.utils 包中(请参考本书附的源代码)。

【例 9-2】 类 BitcoinJController 的代码示例。

```java
package com.bookcode.jcoinapi.controller;
import com.bookcode.jcoinapi.utils.LegacyAddress; //调整了原来的包
import org.bitcoinj.core.*;
import org.bitcoinj.kits.WalletAppKit;
import org.bitcoinj.params.TestNet3Params;
import org.springframework.web.bind.annotation.GetMapping;
import org.springframework.web.bind.annotation.RestController;
import java.io.File;
@RestController
public class BitcoinJController {
    @GetMapping("/testBitcoinj")
    public String blockchain() throws Exception {
        NetworkParameters params = TestNet3Params.get();
        WalletAppKit kit = new WalletAppKit(params, new File("."), "sendrequest-example");
        kit.startAsync();
        kit.awaitRunning();
        LegacyAddress to = LegacyAddress.fromBase58(params,
"mupBAFeT63hXfeeT4rnAUcpKHDkz1n4fdw");
        System.out.println("Send money to: " + to.toString());
```

```
            return "测试 bitcoinj.";
    }
}
```

9.1.4 运行程序

运行程序,在浏览器中输入 localhost:8080/testBitcoinj,在浏览器中输出结果如图 9-1 所示。控制台的主要输出结果如图 9-2 所示。

图 9-1　在浏览器中输入 localhost:8080/testBitcoinj 后浏览器的输出结果

```
[nio-8080-exec-1] org.bitcoinj.core.Context     : Creating bitcoinj 0.14.7 context.
[AppKit STARTING] org.bitcoinj.kits.WalletAppKit : Starting up with directory = .
[AppKit STARTING] org.bitcoinj.core.AbstractBlockChain : chain head is at height
1482379:
 block:
    hash: 0000000000000089c489481ebfff795fe3aff1481a4f175617c8f5d1f4c6276e
    version: 961585152 (BIP34, BIP66, BIP65)
    previous block:
 000000000000001a28bad7cea6162af5ad6c3838e297b431fdbf2271116a58ad
    merkle root:
 037fbbf0f50f60963ee545b4e38c74ff1c91ae6072ebda60f9abbdf9d94d4b9a
    time: 1551404564 (2019-03-01T01:42:44Z)
    difficulty target (nBits): 436307481
    nonce: 3526276356

[eerGroup Thread] org.bitcoinj.core.PeerGroup    : Starting ...
[eerGroup Thread] org.bitcoinj.core.PeerGroup    : Localhost peer not detected.

Send money to: mupBAFeT63hXfeeT4rnAUcpKHDkz1n4fdw
```

图 9-2　在浏览器中输入 localhost:8080/testBitcoinj 后控制台的主要输出结果

9.2　fabric-sdk-java 的应用

9.2.1　fabric-sdk-java 简介

视频讲解

Java SDK for Hyperledger 项目 fabric-sdk-java 有助于促进 Java 应用程序管理 Hyperledger 通道和用户智能合约的生命周期。SDK 提供了在通道上执行用户智能合约、查询块和事务、监视通道上事件的方法。SDK 作用于特定用户的行为,由嵌入式应用程序通过用户界面的实现来定义。SDK 还为 Hyperledger 的证书颁发机构提供客户端。

fabric-sdk-java 包括一些核心类 BlockInfo、BlockchainInfo、TransactionInfo、Peer 等和 exception、helper、security、user 等包。更多信息可以参考 https://github.com/hyperledger/fabric-sdk-java/中的文档和具体代码。

9.2.2 添加依赖

向项目 jcoinapi 的文件 pom.xml 中<dependencies>和</dependencies>之间添加 fabric-sdk-java 依赖,代码如例 9-3 所示。

【例 9-3】 添加 fabric-sdk-java 依赖的代码示例。

```xml
<dependency>
        <groupId>org.hyperledger.fabric-sdk-java</groupId>
        <artifactId>fabric-sdk-java</artifactId>
        <version>1.4.0</version>
</dependency>
```

9.2.3 创建类 HyperledgerController

在包 com.bookcode.jcoinapi.controller 中创建类 HyperledgerController,代码如例 9-4 所示。

【例 9-4】 类 HyperledgerController 的代码示例。

```java
package com.bookcode.jcoinapi.controller;
import java.io.*;
import java.util.*;
import org.hyperledger.fabric.sdk.helper.Config;
import org.hyperledger.fabric.sdk.helper.Utils;
import org.springframework.web.bind.annotation.GetMapping;
import org.springframework.web.bind.annotation.RestController;
@RestController
public class HyperledgerController {
    @GetMapping("/testFabricsdkjava")
    public String blockchain() throws Exception {
        Properties properties = Utils.parseGrpcUrl("grpc://localhost:4");
        String strProp = properties.toString();
        System.out.println("解析 grpc://localhost:4 的结果为" + strProp.toString());
        System.out.println("——————————————————————————");
        Config config = Config.getConfig();
        System.out.println("以下是来自 fabric-sdk-java 中 Config.java 中关于 Crypto 的默认信息:");
```

```java
            System.out.println("————————————————————————————");
            System.out.println(config.getDefaultCryptoSuiteFactory());
            System.out.println(String.valueOf(config.getSecurityLevel()));
            System.out.println(config.getSecurityProviderClassName());
            Map < Integer,String > mc = config.getSecurityCurveMapping();
            System.out.println(config.getHashAlgorithm());
            System.out.println(config.getAsymmetricKeyType());
            System.out.println(config.getCertificateFormat());
            System.out.println(config.getSignatureAlgorithm());
            System.out.println("————————————————————————————");
            System.out.println("以下是来自项目中 config.properties 文件中关于 Crypto 的配置信息:");
            Properties prop = new Properties();
        FileInputStream fis = new FileInputStream("D:\BlockchainProjects\jcoinapi\src\main\resources\config.properties");
            prop.load(fis);
System.out.println(prop.getProperty("org.hyperledger.fabric.sdk.crypto.default_crypto_suite_factory"));
            return "测试 fabric – sdk – java 成功.";
    }
}
```

9.2.4 创建配置文件 config.properties

在目录 src/main/resources 下,创建配置文件 config.properties 并修改配置文件代码,修改后的代码如例 9-5 所示。

【例 9-5】 修改后的配置文件 config.properties 代码示例。

```
org.hyperledger.fabric.sdk.crypto.default_crypto_suite_factory = org.hyperledger.fabric.sdk.security.HLSDKJCryptoSuiteFactory
```

9.2.5 运行程序

运行程序,在浏览器中输入 localhost:8080/testFabricsdkjava,浏览器的输出结果如图 9-3 所示。控制台的输出结果如图 9-4 所示。

图 9-3 在浏览器中输入 localhost:8080/testFabricsdkjava 后浏览器的输出结果

```
解析grpc://localhost:4的结果为{port=4, host=localhost, protocol=grpc}

以下是来自fabric-sdk-java中Config.java中关于Crypto的默认信息:

org.hyperledger.fabric.sdk.security.HLSDKJCryptoSuiteFactory
256
org.bouncycastle.jce.provider.BouncyCastleProvider
SHA2
EC
X.509
SHA256withECDSA

以下是来自项目中config.properties文件中关于Crypto的配置信息:
org.hyperledger.fabric.sdk.security.HLSDKJCryptoSuiteFactory
```

图 9-4　在浏览器中输入 localhost:8080/testFabricsdkjava 后控制台的输出结果

9.3　eth-contract-api 的应用

视频讲解

eth-contract-api 可轻松使用和测试智能合约。

9.3.1　添加依赖

向项目 jcoinapi 的 pom.xml 文件中 < dependencies > 和 </dependencies > 之间添加 eth-contract-api 依赖,代码如例 9-6 所示。

【例 9-6】　添加 eth-contract-api 依赖的代码示例。

```xml
<dependency>
        <groupId>org.adridadou</groupId>
        <artifactId>eth-contract-api</artifactId>
        <version>0.16</version>
</dependency>
```

9.3.2　创建类 EthcontractapiController

在包 com.bookcode.jcoinapi.controller 中创建类 EthcontractapiController,代码如例 9-7 所示。

【例 9-7】　类 EthcontractapiController 的代码示例。

```java
package com.bookcode.jcoinapi.controller;
import org.adridadou.ethereum.ethj.BlockchainConfig;
import org.adridadou.ethereum.ethj.provider.EthereumJConfigs;
import org.adridadou.ethereum.ethj.TestConfig;
import org.springframework.web.bind.annotation.GetMapping;
import org.springframework.web.bind.annotation.RestController;
```

```
import java.util.Date;
@RestController
public class EthcontractapiController {
    @GetMapping("/testEthcontractapi")
    public String blockchain() throws Exception {
        BlockchainConfig.Builder bb = EthereumJConfigs.ropsten();
        System.out.println(bb.eip8(true).build().toString());
        System.out.println( String.valueOf(bb.hashCode()));
        System.out.println( String.valueOf(bb.networkId.id));
        BlockchainConfig blockchainConfig = bb.build();
        System.out.println( blockchainConfig.toString());
        System.out.println("————————————————");
        TestConfig.Builder builder = TestConfig.builder().gasPrice(2);
        System.out.println(builder.toString());
        Date initialTime = new Date();
        System.out.println(builder.initialTime(initialTime).toString());
        return "测试 eth-contract-api 成功.";
    }
}
```

9.3.3 运行程序

运行程序，在浏览器中输入 localhost：8080/testEthcontractapi，浏览器的输出结果如图 9-5 所示。控制台的输出结果如图 9-6 所示。

图 9-5 在浏览器中输入 localhost：8080/testEthcontractapi 后浏览器的输出结果

图 9-6 在浏览器中输入 localhost：8080/testEthcontractapi 后控制器的输出结果

9.4 exonum-java-binding 的应用

视频讲解

exonum-java-binding 是一个用 Java 构建区块链应用程序的框架,它提供了 Exonum 服务运行时环境,允许构建安全的许可区块链应用程序。Exonum 是一个区块链框架,它不是现成的区块链(如 Bitcoin)。Exonum 可用于创建区块链,就像 MVC 框架(如 Struts)可用于创建 Web 应用程序一样。

与 exonum-java-binding 相关的轻客户端 exonum-light-client 是一个与外部 Java 应用程序和 Exonum 区块链交互的库。它可轻松地使用和测试智能合约。更多信息可以参考 https://github.com/exonum 中的文档和具体代码。

9.4.1 添加依赖

向项目 jcoinapi 中 pom.xml 文件的<dependencies>和</dependencies>之间添加 exonum-java-binding 依赖,代码如例 9-8 所示。

【例 9-8】 添加 exonum-java-binding 依赖的代码示例。

```xml
<dependency>
    <groupId>com.exonum.binding</groupId>
    <artifactId>exonum-java-binding-common</artifactId>
    <version>0.4.0</version>
</dependency>
```

9.4.2 创建类 ExonumController

在包 com.bookcode.jcoinapi.controller 中创建类 ExonumController,代码如例 9-9 所示。

【例 9-9】 类 ExonumController 的代码示例。

```java
package com.bookcode.jcoinapi.controller;
import com.exonum.binding.common.hash.HashCode;
import org.springframework.web.bind.annotation.GetMapping;
import org.springframework.web.bind.annotation.RestController;
@RestController
public class ExonumController {
    @GetMapping("/testExonumjavabinding")
    public String blockchain() throws Exception {
        String strPrivateKey = "zhangsansiyao";
        byte [] privateKey = strToByteArray(strPrivateKey);
        HashCode hashCode = HashCode.fromBytes(privateKey);
        String strHash = hashCode.toString();
        System.out.println("Hash Code 是: " + strHash);
        return "测试 exonum-java-binding 成功.";
    }
```

```
    private byte[] strToByteArray(String str) {
        if (str == null) {
            return null;
        }
        byte[] byteArray = str.getBytes();
        return byteArray;
    }
}
```

9.4.3 运行程序

运行程序，在浏览器中输入 localhost:8080/testExonumjavabinding，浏览器的输出结果如图 9-7 所示。控制台的输出如图 9-8 所示。

图 9-7 在浏览器中输入 localhost:8080/testExonumjavabinding 后浏览器的输出结果

Hash Code是：7a68616e6773616e736979616f

图 9-8 在浏览器中输入 localhost:8080/testExonumjavabinding 后控制器的输出结果

9.5 web3j 的应用

视频讲解

9.5.1 web3j 简介

web3j 能够用于处理以太坊智能合约及与以太坊网络上的客户端（节点）进行集成。官方网站上给出的架构图，如图 9-9 所示。该图清楚地说明 web3j 起到了以太坊区块链和 Java 之间的桥梁作用。

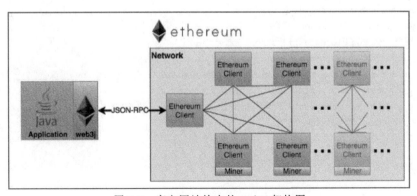

图 9-9 官方网站给出的 web3j 架构图

web3j的特点主要包括以下几方面。

(1) 基于HTTP和IPC的以太坊JSON-RPC客户端API的完整实现。
(2) 对于以太坊钱包的支持。
(3) 自动生成Java智能合约封装包,以创建、部署、交易和调用来自本机Java代码的智能合约(支持Solidity和Truffle定义格式)。
(4) 用于过滤器工作的响应式函数API。

更多信息可以参考 https://github.com/web3j 中的文档和具体代码。

9.5.2 添加依赖

向项目jcoinapi中pom.xml文件的<dependencies>和</dependencies>之间添加web3j依赖,代码如例9-10所示。

【例9-10】 添加web3j依赖的代码示例。

```xml
<dependency>
    <groupId>org.web3j</groupId>
    <artifactId>web3j-spring-boot-starter</artifactId>
    <version>1.6.0</version>
</dependency>
```

9.5.3 创建类Web3jController

在包com.bookcode.jcoinapi.controller中创建类Web3jController,代码如例9-11所示。

【例9-11】 类Web3jController的代码示例。

```java
package com.bookcode.jcoinapi.controller;
import org.springframework.web.bind.annotation.GetMapping;
import org.springframework.web.bind.annotation.RestController;
import org.web3j.crypto.WalletUtils;
import java.io.File;
@RestController
public class Web3jController {
    @GetMapping("/testWeb3j")
    public String blockchain() throws Exception {
        String strDir = WalletUtils.getDefaultKeyDirectory();
        String strFile = WalletUtils.generateNewWalletFile("zs",new File("d:\\"),false);
        System.out.println("默认目录是:" + strDir);
        System.out.println("产生的新文件是:" + strFile);
        return "测试web3j成功.";
    }
}
```

9.5.4　运行程序

运行程序,在浏览器中输入 localhost:8080/testWeb3j,浏览器中的输出结果如图 9-10 所示。控制台中的输出结果如图 9-11 所示,输出默认目录和新产生的 JSON 文件的文件名。新 JSON 文件的内容如图 9-12 所示。

图 9-10　在浏览器中输入 localhost:8080/testWeb3j 后浏览器的输出结果

```
默认目录是：C:\Users\ws\AppData\Roaming\Ethereum
产生的新文件是：UTC--2019-03-02T03-00-08.527000000Z--fe8bb78b974513a212d62a77601022e3559a6989.json
```

图 9-11　在浏览器中输入 localhost:8080/testWeb3j 后控制器的输出结果

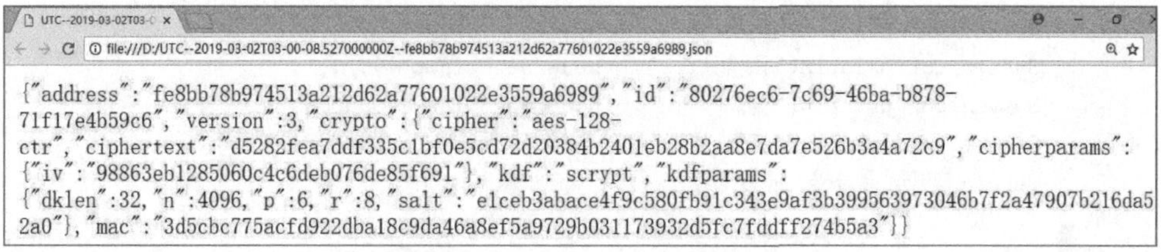

```
{"address":"fe8bb78b974513a212d62a77601022e3559a6989","id":"80276ec6-7c69-46ba-b878-
71f17e4b59c6","version":3,"crypto":{"cipher":"aes-128-
ctr","ciphertext":"d5282fea7ddf335c1bf0e5cd72d20384b2401eb28b2aa8e7da7e526b3a4a72c9","cipherparams":
{"iv":"98863eb1285060c4c6deb076de85f691"},"kdf":"scrypt","kdfparams":
{"dklen":32,"n":4096,"p":6,"r":8,"salt":"e1ceb3abace4f9c580fb91c343e9af3b399563973046b7f2a47907b216da5
2a0"},"mac":"3d5cbc775acfd922dba18c9da46a8ef5a9729b031173932d5fc7fddff274b5a3"}}
```

图 9-12　新 JSON 文件的内容

9.6　WavesJ 的应用

视频讲解

WavesJ 是用于与 Waves 区块链交互的 Java 库。它支持节点交互、离线事务签名、匹配订单以及创建地址和密钥等功能。

9.6.1　添加依赖

向项目 jcoinapi 中 pom.xml 文件的 < dependencies > 和 </dependencies > 之间添加 WavesJ 依赖,代码如例 9-12 所示。

【例 9-12】　添加 WavesJ 依赖的代码示例。

```
< dependency >
        < groupId > com.wavesplatform </groupId >
        < artifactId > wavesj </artifactId >
        < version > 0.13.1 </version >
</dependency >
```

9.6.2 创建类 WavesJController

在包 com.bookcode.jcoinapi.controller 中创建类 WavesJController，代码如例 9-13 所示。

【例 9-13】 类 WavesJController 的代码示例。

```java
package com.bookcode.jcoinapi.controller;
import org.springframework.web.bind.annotation.GetMapping;
import org.springframework.web.bind.annotation.RestController;
import com.wavesplatform.wavesj.*;
@RestController
public class WavesJController {
    @GetMapping("/testWavesJ")
    public String blockchain() throws Exception {
String seed = "health lazy lens fix dwarf salad breeze myself silly december endless rent faculty report beyond";
        PrivateKeyAccount account = PrivateKeyAccount.fromSeed(seed, 0, Account.TESTNET);
        String address = account.getAddress();
        Node node = new Node("https://testnode2.wavesnodes.com/", Account.TESTNET);
        System.out.println("地址是: " + node.getHeight());
        System.out.println("顶点是: " + node.getHeight());
        System.out.println("余额是: " + node.getBalance(address));
        System.out.println("证实是: " + node.getBalance(address, 100));
        return "测试 WavesJ 成功.";
    }
}
```

9.6.3 运行程序

运行程序，在浏览器中输入 localhost:8080/testWavesJ，浏览器的输出结果如图 9-13 所示。控制台的输出结果如图 9-14 所示。

图 9-13 在浏览器中输入 localhost:8080/testWavesJ 后浏览器的输出结果

图 9-14 在浏览器中输入 localhost:8080/testWavesJ 后控制器的输出结果

9.7 基于 web3j 钱包业务功能的示例

视频讲解

本节介绍如何实现基于 web3j 的钱包业务功能。

9.7.1 创建项目并添加依赖

用 IDEA 创建完项目 extx 之后，确保在文件 pom.xml 的 < dependencies > 和 </dependencies > 之间添加了 web3j、Web 等依赖，代码如例 9-14 所示。

【例 9-14】 添加 web3j、Web 等依赖的代码示例。

```
< dependency >
        < groupId > org.web3j </groupId >
        < artifactId > web3j - spring - boot - starter </artifactId >
        < version > 1.6.0 </version >
</dependency >
< dependency >
        < groupId > org.springframework.boot </groupId >
        < artifactId > spring - boot - starter - web </artifactId >
</dependency >
< dependency >
        < groupId > org.springframework.boot </groupId >
        < artifactId > spring - boot - starter - data - jpa </artifactId >
</dependency >
< dependency >
        < groupId > org.springframework.boot </groupId >
        < artifactId > spring - boot - starter - thymeleaf </artifactId >
</dependency >
< dependency >
        < groupId > mysql </groupId >
        < artifactId > mysql - connector - java </artifactId >
</dependency >
< dependency >
        < groupId > org.projectlombok </groupId >
        < artifactId > lombok </artifactId >
        < optional > true </optional >
</dependency >
```

9.7.2 创建类 BlockchainTransaction

在包 com.bookcode 中创建 entity 子包，并在包 com.bookcode.entity 中创建类 BlockchainTransaction，代码如例 9-15 所示。

【例 9-15】 类 BlockchainTransaction 的代码示例。

```
package com.bookcode.entity;
import lombok.Data;
import lombok.NoArgsConstructor;
import javax.persistence.Entity;
import javax.persistence.Id;
import javax.persistence.Table;
@Entity
@Table(name = "ledgers")
@NoArgsConstructor
@Data
public class BlockchainTransaction {
    @Id
    private String id;
    private int fromId;
    private int toId;
    private long value;
    private boolean accepted;
    public BlockchainTransaction(int fromId, int toId, long value) {
        this.fromId = fromId;
        this.toId = toId;
        this.value = value;
    }
}
```

9.7.3 创建接口 BTxRepository

在包 com.bookcode 中创建 dao 子包,并在包 com.bookcode.dao 中创建接口 BTxRepository,代码如例 9-16 所示。

【例 9-16】 BTxRepository 的代码示例。

```
package com.bookcode.dao;
import com.bookcode.entity.BlockchainTransaction;
import org.springframework.data.jpa.repository.JpaRepository;
public interface BTxRepository extends JpaRepository<BlockchainTransaction,String> {
}
```

9.7.4 创建类 BlockchainService

在包 com.bookcode 中创建 service 子包,并在包 com.bookcode.service 中创建类 BlockchainService,代码如例 9-17 所示。

【例 9-17】 类 BlockchainService 的代码示例。

```
package com.bookcode.service;
import com.bookcode.entity.BlockchainTransaction;
```

```java
import org.slf4j.Logger;
import org.slf4j.LoggerFactory;
import org.springframework.stereotype.Service;
import org.web3j.protocol.Web3j;
import org.web3j.protocol.core.DefaultBlockParameterName;
import org.web3j.protocol.core.methods.request.Transaction;
import org.web3j.protocol.core.methods.response.*;
import javax.annotation.PostConstruct;
import javax.annotation.Resource;
import java.io.IOException;
import java.math.BigInteger;
@Service
public class BlockchainService {
    @Resource
    Web3j web3j;
    private static final Logger LOGGER = LoggerFactory.getLogger(BlockchainService.class);
    public BlockchainTransaction process(BlockchainTransaction trx) throws IOException {
        EthAccounts accounts = web3j.ethAccounts().send();
        EthGetTransactionCount transactionCount = web3j.ethGetTransactionCount(accounts.getAccounts().get(trx.getFromId()), DefaultBlockParameterName.LATEST).send();
        Transaction transaction = Transaction.createEtherTransaction(
accounts.getAccounts().get(trx.getFromId()), transactionCount.getTransactionCount(),
BigInteger.valueOf(trx.getValue()),
BigInteger.valueOf(21_000), accounts.getAccounts().get(trx.getToId()), BigInteger.valueOf(trx.getValue()));
        EthSendTransaction response = web3j.ethSendTransaction(transaction).send();
        if (response.getError() != null) {
            trx.setAccepted(false);
            LOGGER.info("Tx rejected: {}", response.getError().getMessage());
            return trx;
        }
        trx.setAccepted(true);
        String txHash = response.getTransactionHash();
        LOGGER.info("业务 Hash 值: {}", txHash);
        trx.setId(txHash);
        EthGetTransactionReceipt receipt = web3j.ethGetTransactionReceipt(txHash).send();
        receipt.getTransactionReceipt().ifPresent(transactionReceipt -> LOGGER.info("Tx receipt: {}", transactionReceipt.getCumulativeGasUsed().intValue()));
        return trx;
    }
    @PostConstruct
    public void listen() {
        web3j.transactionObservable().subscribe(tx -> {
LOGGER.info("新的业务: id = {}, block = {}, from = {}, to = {}, value = {}", tx.getHash(), tx.getBlockHash(), tx.getFrom(), tx.getTo(), tx.getValue().intValue());
            try {
                EthCoinbase coinbase = web3j.ethCoinbase().send();
EthGetTransactionCount transactionCount = web3j.ethGetTransactionCount(tx.getFrom(), DefaultBlockParameterName.LATEST).send();
                LOGGER.info("业务记数: {}", transactionCount.getTransactionCount().intValue());
                if (transactionCount.getTransactionCount().intValue() % 10 == 0) {
```

```
            EthGetTransactionCount tc = web3j.ethGetTransactionCount(coinbase.getAddress(),
            DefaultBlockParameterName.LATEST).send();
            Transaction transaction = Transaction.createEtherTransaction(coinbase.getAddress(), tc.
            getTransactionCount(), tx.getValue(), BigInteger.valueOf(21_000), tx.getFrom(), tx.
            getValue());
                        web3j.ethSendTransaction(transaction).send();
                    }
                } catch (IOException e) {
                    LOGGER.error("Error getting transactions", e);
                }
            });
            LOGGER.info("Subscribed");
    }
}
```

9.7.5 创建类 BlockchainController

在包 com.bookcode 中创建 controller 子包,并在包 com.bookcode.controller 中创建类 BlockchainController,代码如例 9-18 所示。

【例 9-18】 类 BlockchainController 的代码示例。

```
package com.bookcode.controller;
import com.bookcode.dao.BTxRepository;
import com.bookcode.entity.BlockchainTransaction;
import com.bookcode.service.BlockchainService;
import org.springframework.beans.factory.annotation.Autowired;
import org.springframework.stereotype.Controller;
import org.springframework.web.bind.annotation.GetMapping;
import org.springframework.web.bind.annotation.PostMapping;
import org.springframework.web.bind.annotation.ResponseBody;
import javax.annotation.Resource;
import java.io.IOException;
import java.util.List;
@Controller
public class BlockchainController {
    @Autowired
    BTxRepository bTxRepository;
    @Resource
    BlockchainService service;
    @GetMapping("add")
    public String addVote() {
        return "add";
    }
    @PostMapping(value = "/ledgers", params = {"fromId", "toId", "value"})
    @ResponseBody
    public String execute(String fromId, String toId, String value) throws IOException {
BlockchainTransaction transaction = new BlockchainTransaction(Integer.parseInt(fromId),
Integer.parseInt(toId), Long.parseLong(value));
        BlockchainTransaction blockchainTransaction = service.process(transaction);
```

```java
        bTxRepository.save(blockchainTransaction);
        return blockchainTransaction.toString();
    }
    @GetMapping("/ledgers")
    @ResponseBody
    public String getBTxs() {
        List<BlockchainTransaction> blockchainTransactionList = bTxRepository.findAll();
        return blockchainTransactionList.toString();
    }
}
```

9.7.6 创建文件 index.html

在目录 src/main/resources/templates 下,创建文件 index.html,如例 9-19 所示。

【例 9-19】 文件 index.html 的代码示例。

```html
<!DOCTYPE html>
<html xmlns:th="http://www.w3.org/1999/xhtml">
<head>
    <title>增加业务页面</title>
    <meta http-equiv="Content-Type" content="text/html; charset=UTF-8">
</head>
<body>
<h3>业务信息</h3>
<form name="add" th:action="@{/ledgers}" method="post">
    发送者:<input type="text" name="fromId" th:value=""${fromId}"><br/>
    接收者:<input type="text" name="toId" th:value=""${toId}"><br/>
    比特币数值:<input type="text" name="value" th:value=""${value}"><br/>
    <input type="submit" value="增加" />
</form>
</body>
</html>
```

9.7.7 修改配置文件 application.properties

修改配置文件 application.properties,修改后的代码如例 9-20 所示。

【例 9-20】 修改后的配置文件 application.properties 代码示例。

```
spring.datasource.driver-class-name=com.mysql.cj.jdbc.Driver
spring.datasource.username=root
spring.datasource.password=sa
spring.jpa.hibernate.ddl-auto=update
spring.datasource.url=jdbc:mysql://localhost:3306/testnew?serverTimezone=GMT%2B8&useUnicode=true&characterEncoding=UTF-8&useSSL=false
spring.jpa.database-platform:org.hibernate.dialect.MySQL5InnoDBDialect
spring.application.name=transaction-service
web3j.client-address=http://127.0.0.1:8545
```

9.7.8 运行程序

双击 C:\Users\ws\AppData\Roaming\npm（请改成用户自己的安装目录）下的 ganache-cli.cmd 文件，启动 Ganache CLI 工具。

运行程序，在浏览器中输入 localhost:8080/add，后输入发送者、接收者、比特币数值等信息，如图 9-15 所示。单击"增加"按钮，浏览器返回刚输入的相关业务信息，结果如图 9-16 所示。在浏览器中输入 localhost:8080/ledgers 后，浏览器的输出结果如图 9-17 所示。

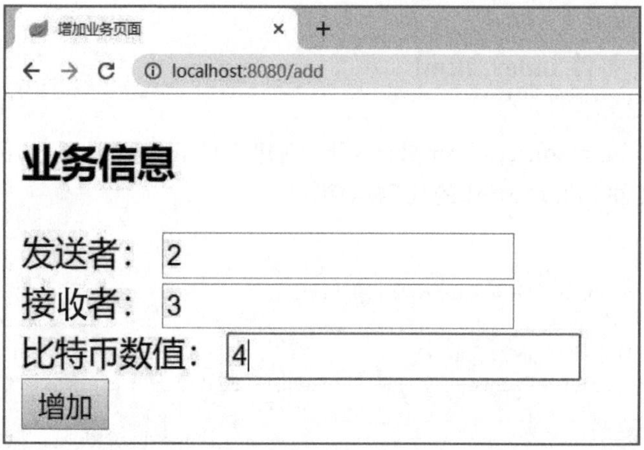

图 9-15 在浏览器中输入 localhost:8080/add 后添加相关信息的结果

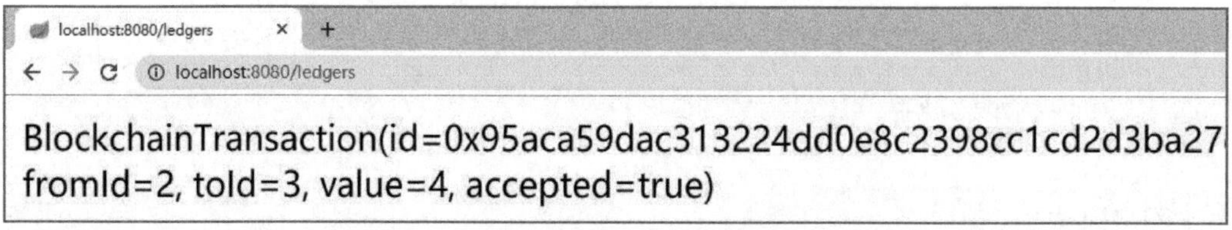

图 9-16 增加一条记录后浏览器的输出结果

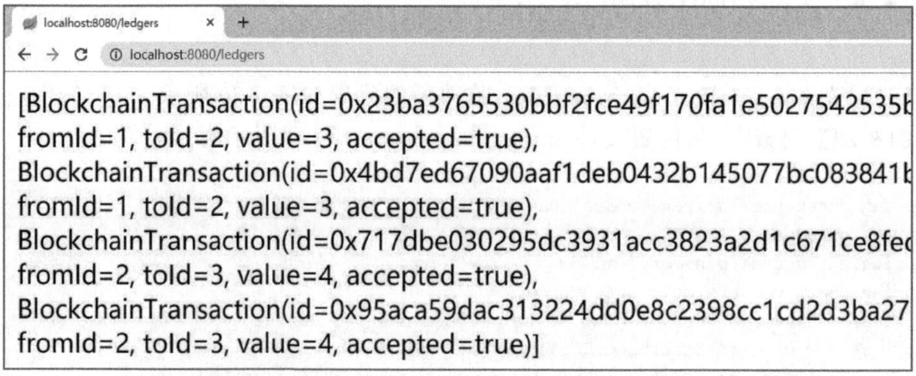

图 9-17 在浏览器中输入 localhost:8080/ledgers 后浏览器的输出结果

习题 9

实验题

1. 实现对 bitcoinj 的应用。
2. 实现对 fabric-sdk-java 的应用。
3. 实现对 eth-contract-api 的应用。
4. 实现对 exonum-java-binding 的应用。
5. 实现对 web3j 的应用。
6. 实现对 WavesJ 的应用。

第10章

基于区块链的简易系统开发

本章介绍三个基于区块链技术的简单应用的开发。

10.1 基于区块链的简易聊天室开发

视频讲解

10.1.1 操作界面

在浏览器中输入 localhost:8080 进行登录,能够输入用户代理名,如图 10-1 所示。首次登录成功后在浏览器中显示的是聊天室基本信息,如图 10-2 所示。已有用户登录聊天室时,新用户登录聊天室之后则会显示已有的聊天信息,如图 10-3 所示。登录成功后可以输入聊天信息,如图 10-4 所示。聊天过程中可以看到对方的聊天信息,如图 10-5 所示。聊天过程中控制台输出对应区块链信息,如图 10-6 所示。

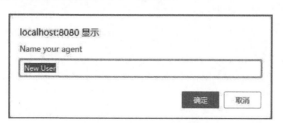

图 10-1 在浏览器中输入 localhost:8080 登录时显示的信息

10.1.2 项目的主要文件构成

主要的文件构成如图 10-7 所示。Block 是区块信息类;Message 是信息类,信息与区块相关。Agent 类是聊天室的用户代理,它处理区块和区块链的信息。AgentManager 类对用户进行增、删、改、查等操作管理，并且在创建 Agent 时为其创建创世块。AgentServerThread 是为了实现多个 Agent 进行聊天的线程问题。ChatController 是相关 URL 操作的映射类。index.html 是默认首页，display.js、restClient.js 和 main.css 是 index.html 文件辅助文件，便于更好地设计和显示。

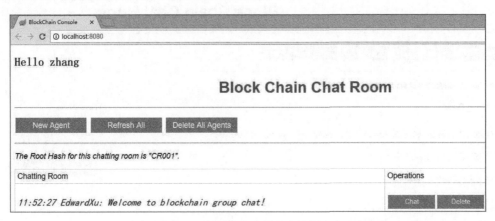

图 10-2　首次登录后在浏览器中显示的聊天室基本信息

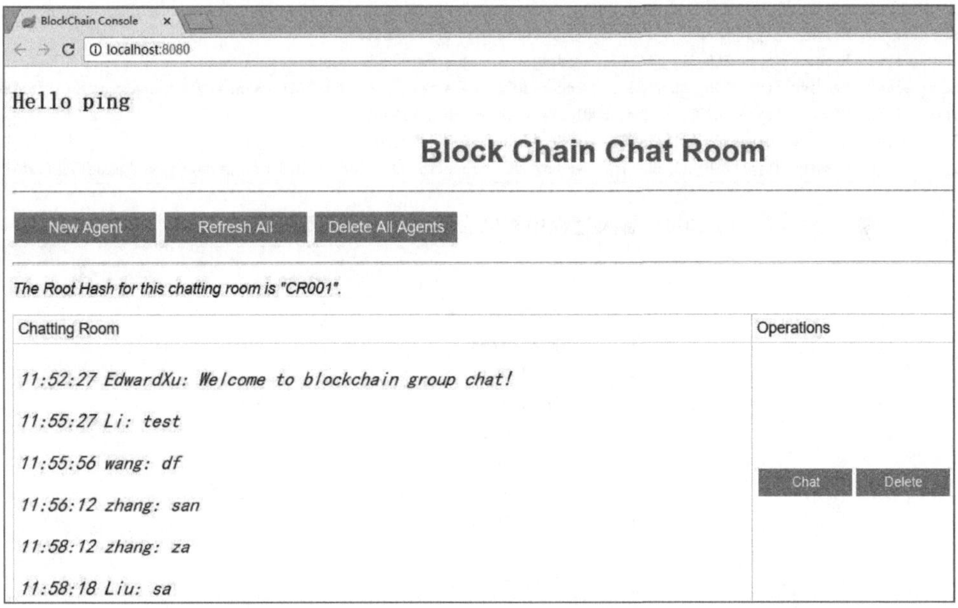

图 10-3　已有用户登录聊天室时新用户登录聊天室之后显示已有的聊天信息

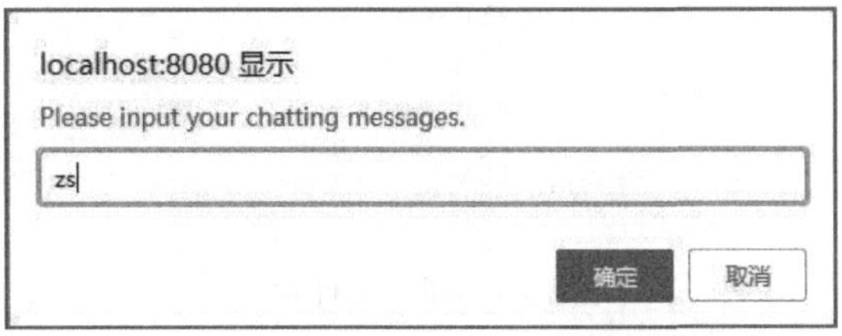

图 10-4　登录成功后可以输入聊天信息

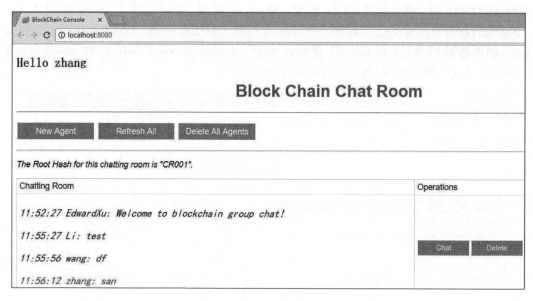

图 10-5 聊天过程中可以看到对方的聊天信息

```
3001 received: Message {type=READY, sender=3001, receiver=0, blocks=null}
3001 received: Message {type=REQ_ALL_BLOCKS, sender=3001, receiver=3001, blocks=[null]}
3001 received: Message {type=RSP_ALL_BLOCKS, sender=3001, receiver=0, blocks=[Block{index=0, timestamp=1551579579536,
3001 received: Message {type=READY, sender=3001, receiver=0, blocks=null}
3001 received: Message {type=REQ_ALL_BLOCKS, sender=3001, receiver=3001, blocks=[null]}
3001 received: Message {type=RSP_ALL_BLOCKS, sender=3001, receiver=0, blocks=[Block{index=0, timestamp=1551579579536,
```

图 10-6 聊天过程中控制台输出的对应区块链信息

图 10-7 项目文件构成

10.1.3 创建项目并添加依赖

用 IDEA 创建完项目 chat 之后,确保在文件 pom.xml 中 <dependencies> 和 </dependencies> 之间添加了 Web 依赖,代码如例 10-1 所示。

【例 10-1】 添加 Web 依赖的代码示例。

```
<dependency>
        <groupId>org.springframework.boot</groupId>
        <artifactId>spring-boot-starter-web</artifactId>
</dependency>
```

10.1.4 创建类 Block

在包 com.bookcode 中创建 entity 子包,并在包 com.bookcode.entity 中创建类 Block,代码如例 10-2 所示。

【例 10-2】 类 Block 的代码示例。

```java
package com.bookcode.entity;
import java.io.Serializable;
import java.security.MessageDigest;
import java.security.NoSuchAlgorithmException;
import java.text.SimpleDateFormat;
import java.util.Date;
public class Block implements Serializable {
    private static final long serialVersionUID = 1L;
    private int index;
    private Long timestamp;
    private String hash;
    private String previousHash;
    private String creator;
    private String message;
    public Block() {
    }
    @Override
    public String toString() {
        return "Block{" +
                "index = " + index +
                ", timestamp = " + timestamp +
                ", creator = " + creator +
                ", hash = '" + hash + '\'' +
                ", previousHash = '" + previousHash + '\'' +
                ", message = '" + message + '\'' +
                '}';
    }
    @Override
```

```java
    public boolean equals(final Object o) {
        if (this == o) {
            return true;
        }
        if (o == null || getClass() != o.getClass()) {
            return false;
        }
        final Block block = (Block) o;
        return index == block.index
                && timestamp.equals(block.timestamp)
                && hash.equals(block.hash)
                && previousHash.equals(block.previousHash)
                && creator.equals(block.creator);
    }
    @Override
    public int hashCode() {
        int result = index;
        result = 31 * result + timestamp.hashCode();
        result = 31 * result + hash.hashCode();
        result = 31 * result + previousHash.hashCode();
        result = 31 * result + creator.hashCode();
        return result;
    }
    public Block(int index, String preHash, String creator, String message) {
        this.index = index;
        this.previousHash = preHash;
        this.creator = creator;
        this.message = message;
        timestamp = System.currentTimeMillis();
        hash = calculateHash(String.valueOf(index) + previousHash + String.valueOf(timestamp));
    }
    public String getCreator() {
        return creator;
    }
    public int getIndex() {
        return index;
    }
    public String getTimestamp() {
        SimpleDateFormat sdf = new SimpleDateFormat("HH:mm:ss");
        Date time = new Date(timestamp);
        return sdf.format(time);
    }
    public String getHash() {
        return hash;
    }
    public String getPreviousHash() {
        return previousHash;
    }
    public String getMessage() {
```

```java
            return message;
    }
    public void setMessage(String message){
        this.message = message;
    }
    private String calculateHash(String text) {
        MessageDigest digest;
        try {
            digest = MessageDigest.getInstance("SHA-256");
        } catch (NoSuchAlgorithmException e) {
            return "HASH_ERROR";
        }
        final byte bytes[] = digest.digest(text.getBytes());
        final StringBuilder hexString = new StringBuilder();
        for (final byte b : bytes) {
            String hex = Integer.toHexString(0xff & b);
            if (hex.length() == 1) {
                hexString.append('0');
            }
            hexString.append(hex);
        }
        return hexString.toString();
    }
}
```

10.1.5　创建类 Agent

在包 com.bookcode.entity 中创建类 Agent,代码如例 10-3 所示。

【例 10-3】 类 Agent 的代码示例。

```java
package com.bookcode.entity;
import java.io.IOException;
import java.io.ObjectInputStream;
import java.io.ObjectOutputStream;
import java.net.ServerSocket;
import java.net.Socket;
import java.net.UnknownHostException;
import java.util.ArrayList;
import java.util.Arrays;
import java.util.List;
import java.util.Objects;
import java.util.concurrent.ScheduledThreadPoolExecutor;
import static com.bookcode.entity.Message.MESSAGE_TYPE.*;
public class Agent {
    private String name;
    private String address;
    private int port;
```

```java
    private List<Agent> peers;
    private List<Block> blockchain = new ArrayList<>();
    private ServerSocket serverSocket;
    private ScheduledThreadPoolExecutor executor = new ScheduledThreadPoolExecutor(10);
    private boolean listening = true;
    public Agent() {
    }
Agent(final String name, final String address, final int port, final Block root, final List<Agent> agents) {
        this.name = name;
        this.address = address;
        this.port = port;
        this.peers = agents;
        blockchain.add(root);
    }
    public String getName() {
        return name;
    }
    public String getAddress() {
        return address;
    }
    public int getPort() {
        return port;
    }
    public List<Block> getBlockchain() {
        return blockchain;
    }
    Block createBlock(String message) {
        if (blockchain.isEmpty()) {
            return null;
        }
        Block previousBlock = getLatestBlock();
        if (previousBlock == null) {
            return null;
        }
        final int index = previousBlock.getIndex() + 1;
        final Block block = new Block(index, previousBlock.getHash(), name, message);
        System.out.println(String.format("%s created new block %s", name, block.toString()));
        broadcast(INFO_NEW_BLOCK, block);
        return block;
    }
    void addBlock(Block block) {
        if (isBlockValid(block)) {
            blockchain.add(block);
        }
    }
    void startHost() {
        executor.execute(() -> {
            try {
```

```java
                serverSocket = new ServerSocket(port);
                    System.out.println(String.format("Server %s started", serverSocket.getLocalPort()));
                listening = true;
                while (listening) {
final AgentServerThread thread = new AgentServerThread(Agent.this, serverSocket.accept());
                    thread.start();
                }
                serverSocket.close();
            } catch (IOException e) {
                System.err.println("Could not listen to port " + port);
            }
        });
        broadcast(REQ_ALL_BLOCKS, null);
    }
    void stopHost() {
        listening = false;
        try {
            serverSocket.close();
        } catch (IOException e) {
            e.printStackTrace();
        }
    }
    private Block getLatestBlock() {
        if (blockchain.isEmpty()) {
            return null;
        }
        return blockchain.get(blockchain.size() - 1);
    }
    private boolean isBlockValid(final Block block) {
        final Block latestBlock = getLatestBlock();
        if (latestBlock == null) {
            return false;
        }
        final int expected = latestBlock.getIndex() + 1;
        if (block.getIndex() != expected) {
System.out.println(String.format("Invalid index. Expected: %s Actual: %s", expected, block.getIndex()));
            return false;
        }
        if (!Objects.equals(block.getPreviousHash(), latestBlock.getHash())) {
            System.out.println("Unmatched hash code");
            return false;
        }
        return true;
    }
    private void broadcast(Message.MESSAGE_TYPE type, final Block block) {
        peers.forEach(peer -> sendMessage(type, peer.getAddress(), peer.getPort(), block));
    }
```

```java
        private void sendMessage(Message.MESSAGE_TYPE type, String host, int port, Block...
blocks) {
            try (
                final Socket peer = new Socket(host, port);
                final ObjectOutputStream out = new ObjectOutputStream(peer.
getOutputStream());
                final ObjectInputStream in = new ObjectInputStream(peer.
getInputStream())) {
                Object fromPeer;
                while ((fromPeer = in.readObject()) != null) {
                    if (fromPeer instanceof Message) {
                        final Message msg = (Message) fromPeer;
                        System.out.println(String.format("%d received: %s", this.port, msg.
toString()));
                        if (READY == msg.type) {
                            out.writeObject(new Message.MessageBuilder()
                                    .withType(type)
                                    .withReceiver(port)
                                    .withSender(this.port)
                                    .withBlocks(Arrays.asList(blocks)).build());
                        } else if (RSP_ALL_BLOCKS == msg.type) {
                            if (!msg.blocks.isEmpty() && this.blockchain.size() == 1) {
                                blockchain = new ArrayList<>(msg.blocks);
                            }
                            break;
                        }
                    }
                }
            } catch (UnknownHostException e) {
                System.err.println(String.format("Unknown host %s %d", host, port));
            } catch (IOException e) {
System.err.println(String.format("%s couldn't get I/O for the connection to %s.
Retrying...%n", getPort(), port));
                try {
                    Thread.sleep(100);
                } catch (InterruptedException e1) {
                    e1.printStackTrace();
                }
            } catch (ClassNotFoundException e) {
                e.printStackTrace();
            }
        }
    }
```

10.1.6 创建类 AgentServerThread

在包 com.bookcode.entity 中创建类 AgentServerThread,代码如例 10-4 所示。

【例 10-4】 类 AgentServerThread 的代码示例。

```java
package com.bookcode.entity;
import java.io.IOException;
import java.io.ObjectInputStream;
import java.io.ObjectOutputStream;
import java.net.Socket;
import static com.bookcode.entity.Message.MESSAGE_TYPE.*;
public class AgentServerThread extends Thread {
    private Socket client;
    private final Agent agent;
    AgentServerThread(final Agent agent, final Socket client) {
        super(agent.getName() + System.currentTimeMillis());
        this.agent = agent;
        this.client = client;
    }
    @Override
    public void run() {
        try (
                ObjectOutputStream out = new ObjectOutputStream(client.getOutputStream());
                final ObjectInputStream in = new ObjectInputStream(client.getInputStream())) {
            Message message = new Message.MessageBuilder().withSender(agent.getPort()).withType(READY).build();
            out.writeObject(message);
            Object fromClient;
            while ((fromClient = in.readObject()) != null) {
                if (fromClient instanceof Message) {
                    final Message msg = (Message) fromClient;
                    System.out.println(String.format("%d received: %s", agent.getPort(), fromClient.toString()));
                    if (INFO_NEW_BLOCK == msg.type) {
                        if (msg.blocks.isEmpty() || msg.blocks.size() > 1) {
                            System.err.println("Invalid block received: " + msg.blocks);
                        }
                        synchronized (agent) {
                            agent.addBlock(msg.blocks.get(0));
                        }
                        break;
                    } else if (REQ_ALL_BLOCKS == msg.type) {
                        out.writeObject(new Message.MessageBuilder()
                                .withSender(agent.getPort())
                                .withType(RSP_ALL_BLOCKS)
                                .withBlocks(agent.getBlockchain())
                                .build());
                        break;
                    }
                }
            }
            client.close();
        } catch (ClassNotFoundException | IOException e) {
            e.printStackTrace();
```

        }
    }
}

10.1.7 创建类 AgentManager

在包 com.bookcode.entity 中创建类 AgentManager,代码如例 10-5 所示。

【例 10-5】 类 AgentManager 的代码示例。

```java
package com.bookcode.entity;
import java.util.ArrayList;
import java.util.List;
public class AgentManager {
    private List<Agent> agents = new ArrayList<>();
    private static final Block root = new Block(0, "CR001", "EdwardXu", "Welcome to blockchain group chat!");
    public Agent addAgent(String name, int port) {
        Agent a = new Agent(name, "localhost", port, root, agents);
        a.startHost();
        agents.add(a);
        return a;
    }
    public Agent getAgent(String name) {
        for (Agent a : agents) {
            if (a.getName().equals(name)) {
                return a;
            }
        }
        return null;
    }
    public List<Agent> getAllAgents() {
        return agents;
    }
    public void deleteAgent(String name) {
        final Agent a = getAgent(name);
        if (a != null) {
            a.stopHost();
            agents.remove(a);
        }
    }
    public List<Block> getAgentBlockchain(String name) {
        final Agent agent = getAgent(name);
        if (agent != null) {
            return agent.getBlockchain();
        }
        return null;
    }
```

```java
    public void deleteAllAgents() {
        for (Agent a : agents) {
            a.stopHost();
        }
        agents.clear();
    }
    public Block createBlock(final String name, final String message) {
        final Agent agent = getAgent(name);
        if (agent != null) {
            return agent.createBlock(message);
        }
        return null;
    }
}
```

10.1.8 创建类 Message

在包 com.bookcode.entity 中创建类 Message,代码如例 10-6 所示。

【例 10-6】 类 Message 的代码示例。

```java
package com.bookcode.entity;
import java.io.Serializable;
import java.util.List;
public class Message implements Serializable {
    private static final long serialVersionUID = 1L;
    int sender;
    int receiver;
    MESSAGE_TYPE type;
    List<Block> blocks;
    public enum MESSAGE_TYPE {
        READY, INFO_NEW_BLOCK, REQ_ALL_BLOCKS, RSP_ALL_BLOCKS
    }
    @Override
    public String toString() {
        return String.format("Message {type = %s, sender = %d, receiver = %d, blocks = %s}", type, sender, receiver, blocks);
    }
    static class MessageBuilder {
        private final Message message = new Message();
        MessageBuilder withSender(final int sender) {
            message.sender = sender;
            return this;
        }
        MessageBuilder withReceiver(final int receiver) {
            message.receiver = receiver;
            return this;
        }
```

```java
            MessageBuilder withType(final MESSAGE_TYPE type) {
                message.type = type;
                return this;
            }
            MessageBuilder withBlocks(final List<Block> blocks) {
                message.blocks = blocks;
                return this;
            }
            Message build() {
                return message;
            }
        }
    }
```

10.1.9 创建类 ChatController

在包 com.bookcode 中创建 controller 子包,并在包 com.bookcode.controller 中创建类 ChatController,代码如例 10-7 所示。

【例 10-7】 类 ChatController 的代码示例。

```java
package com.bookcode.controller;
import com.bookcode.entity.Agent;
import com.bookcode.entity.AgentManager;
import com.bookcode.entity.Block;
import org.springframework.web.bind.annotation.*;
import java.util.List;
@RestController
@RequestMapping("/chat")
public class ChatController {
    private static AgentManager agentManager = new AgentManager();
    @GetMapping("/{name}")
    public Agent getAgent(@PathVariable("name") String name) {
        return agentManager.getAgent(name);
    }
    @DeleteMapping("/")
    public void deleteAgent(@RequestParam("name") String name) {
        agentManager.deleteAgent(name);
    }
    @PostMapping("/{name}/{port}")
    public Agent addAgent(@PathVariable("name") String name, @PathVariable("port") int port) {
        return agentManager.addAgent(name, port);
    }
    @GetMapping("/all")
    public List<Agent> getAllAgents() {
        return agentManager.getAllAgents();
    }
```

```
    @DeleteMapping("/all")
    public void deleteAllAgents() {
        agentManager.deleteAllAgents();
    }
    @PostMapping("/mine/{name}/{message}")
    public Block createBlock(@PathVariable("name") final String name, @PathVariable
("message") final String message) {
        return agentManager.createBlock(name, message);
    }
}
```

10.1.10 创建文件 index.html

在目录 src/main/resources/static 下,创建文件 index.html,代码如例 10-8 所示。

【例 10-8】 文件 index.html 的代码示例。

```html
<!doctype html>
<html>
<head>
    <title>BlockChain Console</title>
    <link rel="stylesheet" href="css/main.css">
</head>
<script src="js/display.js"></script>
<script src="js/restClient.js"></script>
<h2>
<script type="text/javascript">
    var name = prompt("Name your agent","New User")
    if (name!==null && name!=="") {
        document.write("Hello " + name)
    }
    var idx = getNextCount();
    var platformName = navigator.platform;
    var port = 3000 + idx;
    sendHttpRequest("POST", "chat/" + name + "/" + port, null, displayAgent);
</script>
</h2>
<body onload="getAllAgents()">
<script src="js/display.js"></script>
<script src="js/restClient.js"></script>
<h1>Block Chain Chat Room</h1>
<hr>
<p>
<button type="button" class="button" id="addAgent" onclick="createAgent()">New Agent
</button>
    <button type="button" id="getAll" onclick="getAllAgents()" class="button">Refresh
All</button>
```

```
<button type = "button" class = "button" id = "delAllAgent" onclick = "deleteAllAgents()">
Delete All Agents </button>
    <script language = "javascript">
        setInterval("getAllAgents()", 50000);
    </script>
</p>
<hr>
<p id = "msg" class = "msg">
    The Root Hash for this chatting room is "CR001".
</p>
<table id = "output">
</table>
</body>
</html>
```

10.1.11 创建文件 display.js

在目录 src/main/resources/static 下,创建 js 子目录,并在目录 src/main/resources/static/js 下创建文件 display.js,代码如例 10-9 所示。

【例 10-9】 修改后的文件 display.js 代码示例。

```
"use strict";
const OUTPUT_TABLE_NAME = "output";
function displayMsg(text, color) {
    if (color === undefined) {
        color = "black";
    }
    document.getElementById("msg").innerHTML = text;
    document.getElementById("msg").style.color = color;
}
function displayAllAgents(json) {
    cleanTable(OUTPUT_TABLE_NAME);
    var agents;
    try {
        agents = JSON.parse(json);
    } catch (e) {
        displayMsg("Invalid response from server " + json, "red");
        return;
    }
    for (var i in agents){
        if (agents[i].name === name){
            displayAgent(agents[i]);
            break;
        }
    }
}
function displayAgent(jsonAgent) {
```

```javascript
        if (typeof jsonAgent === "string") {
            var agent;
            try {
                agent = JSON.parse(jsonAgent);
            } catch (e) {
                displayMsg("Invalid response from server " + jsonAgent, "red");
                return;
            }
        } else {
            agent = jsonAgent;
        }
        var idx = 0;
        var table = document.getElementById(OUTPUT_TABLE_NAME);
        var row = table.insertRow(table.length);
        const chain = agent.blockchain;
        const blockchainCell = row.insertCell(idx++);
        for (var i in chain) {
            blockchainCell.appendChild(createBlockP(chain[i]));
        }
        blockchainCell.className = "blockchain";
        var p = document.createElement("P");
        p.appendChild(addCellButton("Chat", function () {
            mine(agent.name);
        }));
        p.appendChild(addCellButton("Delete", function () {
            deleteAgent(agent.name);
        }));
        row.insertCell(idx).appendChild(p);
        function addCellButton(name, onclick) {
            var button = document.createElement("BUTTON");
            button.className = "cellButton";
            button.appendChild(document.createTextNode(name));
            button.onclick = onclick;
            return button;
        }
}
function displayBlock(jsonBlock) {
    if (typeof jsonBlock === "string") {
        var block;
        try {
            block = JSON.parse(jsonBlock);
        } catch (e) {
document.getElementById("msg").innerHTML = "Invalid response from server " + jsonBlock;
            return;
        }
    } else {
        block = jsonBlock;
    }
    displayMsg("New block mined:< br >" + getBlockString(block), "green");
}
```

```javascript
function getBlockString(block) {
    return "index = " + block.index + " creator = " + block.creator + " timestamp = "
        + block.timestamp + " message " + block.message + "<br>";
}
function createBlockP(block) {
    var p = document.createElement("P");
    if (block.creator === name)
        p.style.color = "green";
    else
        p.style.color = "black";
    p.title = "creator " + block.creator;
    p.innerHTML = block.timestamp + " " + block.creator + ": " + block.message + "<br>";
    console.log("create p.innerHTML " + p.innerHTML);
    return p;
}
function cleanTable(name) {
    var table = document.getElementById(name);
    table.innerHTML = "";
    var row = table.insertRow(0);
    var idx = 0;
    row.insertCell(idx++).innerHTML = "Chatting Room";
    row.insertCell(idx).innerHTML = "Operations";
}
```

10.1.12 创建文件 restClient.js

在目录 src/main/resources/static/js 下创建文件 restClient.js 并修改文件,修改后的代码如例 10-10 所示。

【例 10-10】 修改后的文件 restClient.js 代码示例。

```javascript
"use strict";
function getAllAgents() {
    sendHttpRequest("GET", "chat/all", null, displayAllAgents);
}
function deleteAllAgents() {
    sendHttpRequest("DELETE", "chat/all", null, getAllAgents);
}
function createAgent() {
    var idx = getNextCount();
    var platformName = navigator.platform;
    var name = prompt("Please input your name.", platformName);
    var port = 3000 + idx;
    sendHttpRequest("POST", "chat/" + name + "/" + port, null, displayAgent);
}
function deleteAgent(name) {
    sendHttpRequest("DELETE", "chat/" + name, null, getAllAgents);
}
function getAgent() {
```

```javascript
        var name = document.getElementById("agentNameGet").value;
        sendHttpRequest("GET", "chat/" + name, null, null);
    }
    function mine(name) {
        var platformMessage = navigator.platform;
        var message = prompt("Please input your chatting messages.", platformMessage);
        sendHttpRequest("POST", "chat/mine/" + name + "/" + message, null, getAllAgents);
    }
    function sendHttpRequest(action, url, data, callback) {
        var xmlHttp = new XMLHttpRequest();
        xmlHttp.onreadystatechange = function () {
            if (xmlHttp.readyState === 4 && xmlHttp.status === 200) {
                callback(xmlHttp.responseText);
            }
        };
        xmlHttp.open(action, url, true);
        xmlHttp.send(data);
    }
    var getNextCount = (function () {
        if (!sessionStorage.count) {
            sessionStorage.count = 0;
        }
        return function () {
            sessionStorage.count = Number(sessionStorage.count) + 1;
            return Number(sessionStorage.count);
        }
    })();
```

10.1.13 创建文件 main.css

在目录 src/main/resources/static 下,创建 css 子目录,并在目录 src/main/resources/static/css 下创建文件 main.css,修改文件代码,修改后的代码如例 10-11 所示。

【例 10-11】 修改后的文件 main.css 的代码示例。

```css
button.button:hover {
    color: orangered;
    box-shadow: 0 2px 5px 0 rgba(0, 0, 0, 0.16), 0 2px 10px 0 rgba(0, 0, 0, 0.12);
}
button.button {
    background-color: #4CAF50;
    border: none;
    color: white;
    width: 150px;
    padding: 6px 10px;
    text-align: center;
    text-decoration: none;
    display: inline-block;
```

```css
        font-size: 16px;
        margin: 4px 2px;
        cursor: pointer;
    }
    button.cellButton {
        background-color: #4CAF50;
        border: none;
        color: white;
        width: 100px;
        padding: 6px 10px;
        text-align: center;
        text-decoration: none;
        display: inline-block;
        font-size: 14px;
        margin: 4px 2px;
        cursor: pointer;
    }
    p.agent {
        font-size: 18px;
        font-family: Arial, serif;
        font-weight: bold;
    }
    p.msg {
        font-size: 16px;
        font-style: italic;
        font-family: Arial, serif;
    }
    input {
        font-size: 16px;
        font-family: Arial, fantasy;
    }
    h1 {
        text-align: center;
        color: crimson;
        font-family: Arial, fantasy;
    }
    h2 {
        font-size: 24px;
        color: navy;
        font-family: Bookman L, serif
    }
    table {
        font-family: arial, sans-serif;
        border-collapse: collapse;
        width: 90%;
    }
    td, th {
        border: 1px solid #dddddd;
        text-align: left;
        padding: 5px;
```

```
}
td.blockchain {
    font-size: 20px;
    font-style: italic;
    font-family: URW Chancery L, cursive
}
spring.cloud.nacos.config.server-addr = 127.0.0.1:8848
spring.application.name = example
```

10.1.14 运行程序

运行程序，在浏览器中输入 localhost:8080，结果如图 10-1 所示。首次登录成功后在浏览器中显示的是聊天室基本信息，如图 10-2 所示。

在图 10-2 中单击 New Agent 按钮，在弹出的对话框中可以增加新的用户代理，如图 10-8 所示。单击 Refresh All 按钮就会刷新所有 Agent 信息；单击 Delete All Agents 按钮就删除所有用户代理，需要新建用户代理后才能作为用户代理发送信息。

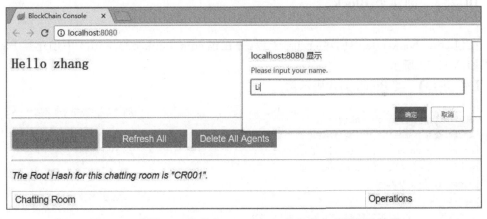

图 10-8　在图 10-2 中单击 New Agent 按钮弹出对话框的结果

10.2　基于区块链的简易证书系统开发

10.2.1　创建项目并添加依赖

用 IDEA 创建完项目 excertificate 之后，确保在文件 pom.xml 中 < dependencies > 和 </dependencies> 之间添加了 Web 等依赖，代码如例 10-12 所示。

【例 10-12】　添加 Web 等依赖的代码示例。

```
<dependency>
    <groupId>org.springframework.boot</groupId>
    <artifactId>spring-boot-starter-web</artifactId>
```

```xml
        </dependency>
        <dependency>
            <groupId>org.springframework.boot</groupId>
            <artifactId>spring-boot-starter-thymeleaf</artifactId>
        </dependency>
        <dependency>
            <groupId>org.springframework.boot</groupId>
            <artifactId>spring-boot-starter-data-jpa</artifactId>
        </dependency>
        <dependency>
            <groupId>mysql</groupId>
            <artifactId>mysql-connector-java</artifactId>
        </dependency>
        <dependency>
            <groupId>org.projectlombok</groupId>
            <artifactId>lombok</artifactId>
        </dependency>
```

10.2.2 创建类 Block

在包 com.bookcode 中创建 entity 子包,并在包 com.bookcode.entity 中创建类 Block,代码如例 10-13 所示。

【例 10-13】 类 Block 的代码示例。

```java
package com.bookcode.entity;
import com.bookcode.utils.PowResult;
import com.bookcode.utils.ProofOfWork;
import lombok.AllArgsConstructor;
import lombok.Data;
import lombok.NoArgsConstructor;
import org.apache.commons.codec.binary.Hex;
import java.time.Instant;
@Data
@AllArgsConstructor
@NoArgsConstructor
public class Block {
    private static final String ZERO_HASH = Hex.encodeHexString(new byte[32]);
    private String hash;
    private String prevBlockHash;
    private String data;
    private long timeStamp;
    private long nonce;
    public static Block newGenesisBlock() {
        return Block.newBlock(ZERO_HASH, "Genesis Block");
    }
    public static Block newBlock(String previousHash, String data) {
        Block block = new Block("", previousHash, data, Instant.now().getEpochSecond(), 0);
```

```
            ProofOfWork pow = ProofOfWork.newProofOfWork(block);
            PowResult powResult = pow.run();
            block.setHash(powResult.getHash());
            block.setNonce(powResult.getNonce());
            return block;
    }
}
```

10.2.3 创建类 Certificate

在包 com.bookcode.entity 中创建类 Certificate,代码如例 10-14 所示。

【例 10-14】 类 Certificate 的代码示例。

```
package com.bookcode.entity;
import lombok.AllArgsConstructor;
import lombok.Data;
import lombok.NoArgsConstructor;
import javax.persistence.*;
@Entity
@Table
@Data
@AllArgsConstructor
@NoArgsConstructor
public class Certificate {
    @Id
    @GeneratedValue(strategy = GenerationType.AUTO)
    @Column
    Long id;
    String studentName;
    String studentSex;
    String studentBirthday;
    String studentBETime;
    String universityName;
    String universityMajor;
    String studentSign;
    String cerType;
    String cerTime;
    String AdminName;
    String AdminNo;
    String Description;
    String BlockchainInfo;
    public Certificate(String studentName, String studentSex, String studentBirthday, String studentBETime, String universityName, String universityMajor, String studentSign, String cerType, String cerTime, String adminName, String adminNo, String description, String blockchainInfo) {
        this.studentName = studentName;
        this.studentSex = studentSex;
```

```
            this.studentBirthday = studentBirthday;
            this.studentBETime = studentBETime;
            this.universityName = universityName;
            this.universityMajor = universityMajor;
            this.studentSign = studentSign;
            this.cerType = cerType;
            this.cerTime = cerTime;
            AdminName = adminName;
            AdminNo = adminNo;
            Description = description;
            BlockchainInfo = blockchainInfo;
    }
}
```

10.2.4　创建接口 CertificateRepository

在包 com.bookcode 中创建 dao 子包,并在包 com.bookcode.dao 中创建接口 CertificateRepository,代码如例 10-15 所示。

【例 10-15】 接口 CertificateRepository 的代码示例。

```
package com.bookcode.dao;
import com.bookcode.entity.Certificate;
import org.springframework.data.jpa.repository.JpaRepository;
public interface CertificateRepository extends JpaRepository<Certificate,Long> {
}
```

10.2.5　创建类 CertificateService

在包 com.bookcode 中创建 service 子包,并在包 com.bookcode.service 中创建类 CertificateService,代码如例 10-16 所示。

【例 10-16】 类 CertificateService 的代码示例。

```
package com.bookcode.service;
import com.bookcode.dao.CertificateRepository;
import com.bookcode.entity.Certificate;
import org.springframework.stereotype.Component;
import javax.annotation.Resource;
import java.util.List;
@Component
public class CertificateService {
    @Resource
    CertificateRepository certificateRepository;
    public Certificate save(Certificate certificate) {
        return certificateRepository.save(certificate);
```

```
    }
    public List<Certificate> findCertificats(Long id){
        return certificateRepository.findAll();
    }
}
```

10.2.6 创建类 ByteUtils

在包 com.bookcode 中创建 utils 子包,并在包 com.bookcode.utils 中创建类 ByteUtils,代码如例 10-17 所示。

【例 10-17】 类 ByteUtils 的代码示例。

```
package com.bookcode.utils;
import org.apache.commons.lang3.ArrayUtils;
import java.nio.ByteBuffer;
import java.util.Arrays;
import java.util.stream.Stream;
public class ByteUtils {
    //将多字节数组合并成一字节数组
    public static byte[] merge(byte[]... bytes) {
        Stream<Byte> stream = Stream.of();
        for (byte[] b : bytes) {
            stream = Stream.concat(stream, Arrays.stream(ArrayUtils.toObject(b)));
        }
        return ArrayUtils.toPrimitive(stream.toArray(Byte[]::new));
    }
    //long 类型转换为 byte[]
    public static byte[] toBytes(long val) {
        return ByteBuffer.allocate(Long.BYTES).putLong(val).array();
    }
}
```

10.2.7 创建类 ProofOfWork

在包 com.bookcode.utils 中创建类 ProofOfWork,代码如例 10-18 所示。

【例 10-18】 类 ProofOfWork 的代码示例。

```
package com.bookcode.utils;
import com.bookcode.entity.Block;
import lombok.Data;
import org.apache.commons.codec.digest.DigestUtils;
import org.apache.commons.lang3.StringUtils;
import java.math.BigInteger;
@Data
```

```java
public class ProofOfWork {
    public static final int TARGET_BITS = 26;
    private Block block;
    private BigInteger target;
    private ProofOfWork(Block block, BigInteger target) {
        this.block = block;
        this.target = target;
    }
    public static ProofOfWork newProofOfWork(Block block) {
        BigInteger targetValue = BigInteger.ONE.shiftLeft((256 - TARGET_BITS));
        return new ProofOfWork(block, targetValue);
    }
    public PowResult run() {
        long nonce = 0;
        String shaHex = "";
        System.out.printf("Mining the block containing: %s \n", this.getBlock().getData());
        long startTime = System.currentTimeMillis();
        while (nonce < Long.MAX_VALUE) {
            byte[] data = this.prepareData(nonce);
            shaHex = DigestUtils.sha256Hex(data);
            if (new BigInteger(shaHex, 16).compareTo(this.target) == -1) {
System.out.printf("Elapsed Time: %s seconds \n", (float) (System.currentTimeMillis() - startTime) / 1000);
                System.out.printf("correct hash Hex: %s \n\n", shaHex);
                break;
            } else {
                nonce++;
            }
        }
        return new PowResult(nonce, shaHex);
    }
    public boolean validate() {
        byte[] data = this.prepareData(this.getBlock().getNonce());
        return new BigInteger(DigestUtils.sha256Hex(data), 16).compareTo(this.target) == -1;
    }
    private byte[] prepareData(long nonce) {
        byte[] prevBlockHashBytes = {};
        if (StringUtils.isNoneBlank(this.getBlock().getPrevBlockHash())) {
prevBlockHashBytes = new BigInteger(this.getBlock().getPrevBlockHash(), 16).toByteArray();
        }
        return ByteUtils.merge(
                prevBlockHashBytes,
                this.getBlock().getData().getBytes(),
                ByteUtils.toBytes(this.getBlock().getTimeStamp()),
                ByteUtils.toBytes(TARGET_BITS),
                ByteUtils.toBytes(nonce)
        );
    }
}
```

10.2.8 创建类 PowResult

在包 com.bookcode.utils 中创建类 PowResult,代码如例 10-19 所示。

【例 10-19】 类 PowResult 的代码示例。

```
package com.bookcode.utils;
import lombok.AllArgsConstructor;
import lombok.Data;
@Data
@AllArgsConstructor
public class PowResult {
    private long nonce;
    private String hash;
}
```

10.2.9 创建类 SearchCertificateController

在包 com.bookcode 中创建 controller 子包,并在包 com.bookcode.controller 中创建类 SearchCertificateController,代码如例 10-20 所示。

【例 10-20】 类 SearchCertificateController 的代码示例。

```
package com.bookcode.controller;
import com.bookcode.entity.Block;
import com.bookcode.entity.Certificate;
import com.bookcode.service.CertificateService;
import org.springframework.beans.factory.annotation.Autowired;
import org.springframework.stereotype.Controller;
import org.springframework.ui.Model;
import org.springframework.web.bind.annotation.GetMapping;
import org.springframework.web.bind.annotation.ModelAttribute;
import org.springframework.web.bind.annotation.PostMapping;
import org.springframework.web.bind.annotation.ResponseBody;
import java.util.ArrayList;
import java.util.List;
@Controller
public class SearchCertificateController {
    @Autowired
    CertificateService certificateService;
    @GetMapping("/certificates")
    public String addcertificate(Model model) {
        model.addAttribute("certificate", new Certificate());
        return "add_certificate";
    }
    @GetMapping("/search")
    public String searchForm(Model model) {
```

```java
        model.addAttribute("certificate", new Certificate());
        return "searchcer";
    }
    @PostMapping("/search")
    @ResponseBody
    public String searchSubmit(@ModelAttribute Certificate certificate) {
        Long id = certificate.getId();
        List<Certificate> certificateList = certificateService.findCertificats(id);
        return certificateList.toString();
    }
    @PostMapping("/certificates")
    @ResponseBody
    public String certificateSubmit(@ModelAttribute Certificate c) {
        List<Block> newblockchain = new ArrayList<Block>();
        newblockchain.add(Block.newGenesisBlock());
        Certificate certificate1 = new Certificate(c.getStudentName(), c.getStudentSex(),
c.getStudentBirthday(), c.getStudentBETime(), c.getUniversityName(), c.getUniversityMajor(),
c.getStudentSign(), c.getCerType(), c.getCerTime(), c.getAdminName(), c.getAdminNo(),
c.getDescription(), newblockchain.toString());
        certificateService.save(certificate1);
        return certificate1.toString();
    }
}
```

10.2.10 创建文件 add_certificate.html

在目录 src/main/resources/templates 下，创建文件 add_certificate.html，代码如例 10-21 所示。

【例 10-21】 文件 add_certificate.html 的代码示例。

```html
<!DOCTYPE html>
<html lang="en" xmlns:th="http://www.w3.org/1999/xhtml">
<head>
    <meta charset="UTF-8">
    <title>输入证书信息页面</title>
</head>
<body>
<h3>证书信息登记</h3>
<form name="addcertificate" th:action="@{/certificates}" th:object="${certificate}" method="post">
    姓名：<input type="text" name="sName" th:field="*{studentName}"><br/>
    性别：<input type="text" name="sSex" th:field="*{studentSex}"><br/>
    出生日期：<input type="text" name="sBirth" th:field="*{studentBirthday}"><br/>
    学习时间：<input type="text" name="sBET" th:field="*{studentBETime}"><br/>
    学校：<input type="text" name="uName" th:field="*{universityName}"><br/>
    专业：<input type="text" name="unMajor" th:field="*{universityMajor}"><br/>
    签发人：<input type="text" name="sSign" th:field="*{studentSign}"><br/>
```

```
证书类型：<input type = "text" name = "cType" th:field = " * {cerType}"><br/>
证书签发时间：<input type = "text" name = "cTime" th:field = " * {cerTime}"><br/>
主管部门：<input type = "text" name = "AName" th:field = " * {AdminName}"><br/>
存档编号：<input type = "text" name = "ANo" th:field = " * {AdminNo}"><br/>
备注：<input type = "text" name = "des" th:field = " * {Description}"><br/>
<input type = "submit" value = "产生证书" />
</form>
</body>
</html>
```

10.2.11 创建文件 searchcer.html

在目录 src/main/resources/templates 下，创建文件 searchcer.html，代码如例 10-22 所示。

【例 10-22】 文件 searchcer.html 的代码示例。

```
<!DOCTYPE html >
< html xmlns:th = "http://www.w3.org/1999/xhtml">
< head >
    <title>证书查询页面</title>
    < meta http - equiv = "Content - Type" content = "text/html; charset = UTF - 8">
</head>
< body >
< div >证书查询</div >
< form name = "search" th:action = "@{/search}" th:object = " $ {certificate}" method = "post" >
    < p >要查询的证书编号：< input type = "text" th:field = " * {id}" /></p>
    < p >< input type = "submit" value = "查询" /> < input type = "reset" value = "清空" /></p>
</form>
</body>
</html>
```

10.2.12 修改配置文件 application.properties

修改配置文件 application.properties，修改后的代码如例 10-23 所示。

【例 10-23】 修改后的配置文件 application.properties 的代码示例。

```
spring.datasource.driver - class - name = com.mysql.cj.jdbc.Driver
spring.datasource.username = root
spring.datasource.password = sa
spring.jpa.hibernate.ddl - auto = update
spring.datasource.url = jdbc:mysql://localhost:3306/testnew?serverTimezone = GMT%2B8&useUnicode = true&characterEncoding = UTF - 8&useSSL = false
spring.jpa.database - platform: org.hibernate.dialect.MySQL5InnoDBDialect
```

10.2.13 运行程序

运行程序，在浏览器中输入 localhost:8080/certificates 后输入证书信息，如图 10-9 所示。单击“产生证书”按钮，控制台输出区块链相关信息，如图 10-10 所示。与此同时，浏览器返回刚输入的证书相关信息，结果如图 10-11 所示。在浏览器中输入 localhost:8080/search 后，输入要查询的证书编号，如图 10-12 所示，单击“查询”按钮，返回证书的相关信息，如图 10-13 所示。

输入证书信息页面
localhost:8080/certificates
证书信息登记
姓名: 赵高
性别: 男
出生日期: 19980101
学习时间: 2015-2019
学校: 秦川大学
专业: 管理学
签发人: 荀子
证书类型: 学士学位
证书签发时间: 20190630
主管部门: 学位办公室
存档编号: XWBQC0002
备注: 教育证书登记
产生证书

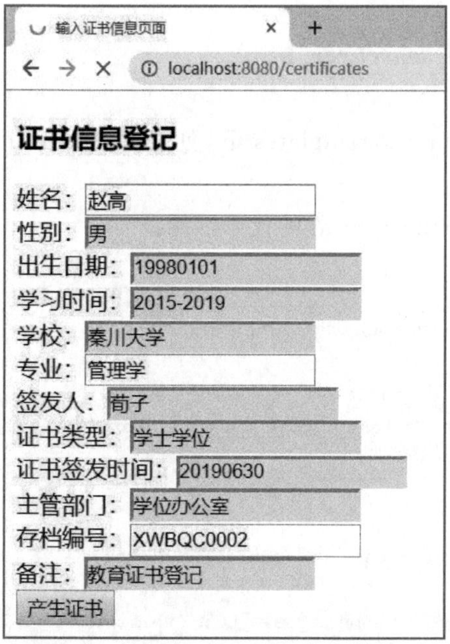

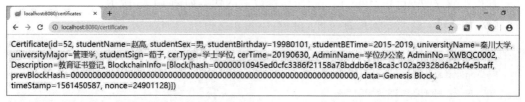

图 10-9 在浏览器中输入 localhost:8080/certificates 后输入证书信息

```
Mining the block containing: Genesis Block
Elapsed Time: 275.978 seconds
correct hash Hex: 00000010945ed0cfc3386f21158a78bddb6e18ca3c102a29328d6a2bf4e5baff
```

图 10-10 在图 10-9 的基础上单击“产生证书”后控制台的输出结果

localhost:8080/certificates
localhost:8080/certificates
Certificate(id=52, studentName=赵高, studentSex=男, studentBirthday=19980101, studentBETime=2015-2019, universityName=秦川大学, universityMajor=管理学, studentSign=荀子, cerType=学士学位, cerTime=20190630, AdminName=学位办公室, AdminNo=XWBQC0002, Description=教育证书登记, BlockchainInfo=[Block(hash=00000010945ed0cfc3386f21158a78bddb6e18ca3c102a29328d6a2bf4e5baff, prevBlockHash=0000000000000000000000000000000000000000000000000000000000000000, data=Genesis Block, timeStamp=1561450587, nonce=24901128)])

图 10-11 在图 10-9 的基础上单击“产生证书”按钮后浏览器的输出结果

第10章　基于区块链的简易系统开发

图10-12　在浏览器中输入localhost:8080/search后输入证书编号

图10-13　在图10-12的基础上单击"查询"按钮后浏览器的输出结果

10.3　基于区块链的简易投票系统开发

视频讲解

10.3.1　创建项目并添加依赖

用 IDEA 创建完项目 exvotesys 之后，确保在文件 pom.xml 中<dependencies>和</dependencies>之间添加了 Web 等依赖，代码如例 10-12 所示。

10.3.2　创建类 ElectionBlock

在包 com.bookcode 中创建 entity 子包，并在包 com.bookcode.entity 中创建类 ElectionBlock，代码如例 10-24 所示。

【例 10-24】　类 ElectionBlock 的代码示例。

```
package com.bookcode.entity;
import lombok.Data;
import lombok.NoArgsConstructor;
import javax.persistence.*;
```

```java
@NoArgsConstructor
@Entity
@Table(name = "electionBlock")
@Data
public class ElectionBlock {
    @GeneratedValue(strategy = GenerationType.AUTO)
    @Id
    @Column
    Integer id;
    public Integer no;
    public String epochTime;
    public String content;
    public String thisBlockHash;
    public String previousBlockHash;
    public ElectionBlock(Integer no, String epochTime, String content, String thisBlockHash, String previousBlockHash) {
        this.no = no;
        this.epochTime = epochTime;
        this.content = content;
        this.thisBlockHash = thisBlockHash;
        this.previousBlockHash = previousBlockHash;
    }
}
```

10.3.3 创建类 Elections

在包 com.bookcode.entity 中创建类 Elections，代码如例 10-25 所示。

【例 10-25】 类 Elections 的代码示例。

```java
package com.bookcode.entity;
import lombok.Data;
import lombok.NoArgsConstructor;
import javax.persistence.*;
@NoArgsConstructor
@Entity
@Table(name = "elections")
@Data
public class Elections {
    @Id
    @GeneratedValue(strategy = GenerationType.AUTO)
    @Column
    Integer id;
    String election_name;
    char active;
    public Elections(String election_name, char active) {
        this.election_name = election_name;
        this.active = active;
```

            }
        }

10.3.4 创建类 Voters

在包 com.bookcode.entity 中创建类 Voters,代码如例 10-26 所示。

【例 10-26】 类 Voters 的代码示例。

```java
package com.bookcode.entity;
import lombok.Data;
import lombok.NoArgsConstructor;
import javax.persistence.*;
@NoArgsConstructor
@Entity
@Table(name = "voters")
@Data
public class Voters {
    @Id
    @GeneratedValue(strategy = GenerationType.AUTO)
    @Column
    Integer id;
    String vname;
    String electionName;
    String public_key;
    String private_key;
    public Voters(String vname, String electionName, String public_key, String private_key) {
        this.vname = vname;
        this.electionName = electionName;
        this.public_key = public_key;
        this.private_key = private_key;
    }
}
```

10.3.5 创建类 Votes

在包 com.bookcode.entity 中创建类 Votes,代码如例 10-27 所示。

【例 10-27】 类 Votes 的代码示例。

```java
package com.bookcode.entity;
import lombok.Data;
import lombok.NoArgsConstructor;
import javax.persistence.*;
@NoArgsConstructor
@Entity
```

```java
@Table(name = "votes")
@Data
public class Votes {
    @Id
    @GeneratedValue(strategy = GenerationType.AUTO)
    @Column
    Integer id;
    String vname;
    String pubkey;
    String prikey;
    String election_name;
    String candicate;
    public Votes(String vname, String pubkey, String prikey, String election_name, String candicate) {
        this.vname = vname;
        this.pubkey = pubkey;
        this.prikey = prikey;
        this.election_name = election_name;
        this.candicate = candicate;
    }
}
```

10.3.6 创建类 Candidates

在包 com.bookcode.entity 中创建类 Candidates，代码如例 10-28 所示。

【例 10-28】 类 Candidates 的代码示例。

```java
package com.bookcode.entity;
import lombok.Data;
import lombok.NoArgsConstructor;
import javax.persistence.*;
@NoArgsConstructor
@Entity
@Table(name = "candidates")
@Data
public class Candidates {
    @Id
    @GeneratedValue(strategy = GenerationType.AUTO)
    @Column
    Integer id;
    String electionName;
    String candidateName;
    public Candidates(String electionName, String candidateName) {
        this.electionName = electionName;
        this.candidateName = candidateName;
    }
}
```

10.3.7 创建实体类访问数据库接口

在包 com.bookcode 中创建 dao 子包,并在包 com.bookcode.dao 中创建接口 ElectionBlockRepository、ElectionsRepository、VotersRepository、VotesRepository、CandidatesRepository。接口 ElectionBlockRepository 的代码如例 10-29 所示。考虑到这些接口语法相同和节约篇幅,其他几个接口的代码示例请参考本书附带的源代码。

【例 10-29】 接口 ElectionBlockRepository 的代码示例。

```java
package com.bookcode.dao;
import com.bookcode.entity.ElectionBlock;
import org.springframework.data.jpa.repository.JpaRepository;
public interface ElectionBlockRepository extends JpaRepository<ElectionBlock, Integer> {
}
```

10.3.8 创建类 HomeController

在包 com.bookcode 中创建 controller 子包,并在包 com.bookcode.controller 中创建类 HomeController,代码如例 10-30 所示。

【例 10-30】 类 HomeController 的代码示例。

```java
package com.bookcode.Controller;
import org.springframework.stereotype.Controller;
import org.springframework.web.bind.annotation.GetMapping;
@Controller
public class HomeController {
    @GetMapping("index")
    public String index(){
        return "index";
    }
    @GetMapping("viewvotes")
    public String viewVotes(){
        return "view_votes";
    }
    @GetMapping("addvote")
    public String addVote(){
        return "add_vote";
    }
    @GetMapping("addvoters")
    public String addVoters(){
        return "add_voters";
    }
    @GetMapping("addcandidates")
    public String addCandidates(){
        return "add_candidates";
```

```
    }
    @GetMapping("addelection")
    public String addEelection(){
        return "add_election";
    }
}
```

10.3.9 创建类 CreatElectionController

在包 com.bookcode.controller 中创建类 CreatElectionController,代码如例 10-31 所示。

【例 10-31】 类 CreatElectionController 的代码示例。

```
package com.bookcode.Controller;
import com.bookcode.dao.*;
import com.bookcode.entity.*;
import org.springframework.beans.factory.annotation.Autowired;
import org.springframework.web.bind.annotation.GetMapping;
import org.springframework.web.bind.annotation.PostMapping;
import org.springframework.web.bind.annotation.RestController;
import java.util.LinkedList;
import java.util.List;
@RestController
public class CreatElectionController {
    @Autowired
    private ElectionsRepository electionsRepository;
    //增加选举
    @PostMapping(value = "/elections", params = {"name"})
    public String addelections(String name) {
        Elections elections = new Elections(name, 'Y');
        electionsRepository.save(elections);
        System.out.println("您发起了一项选举(名称为): " + name + ".");
        return "您发起了一项选举(名称为): " + name + ".";
    }
    @GetMapping(value = "/elections")
    public List<String> queryAllElections() {
        List<String> lists = new LinkedList<String>();
        lists.add("所有选举的信息: ");
        System.out.println("所有选举的信息: ");
        List<Elections> elections = electionsRepository.findAll();
        if (!elections.isEmpty()) {
            for (int i = 0; i < elections.size(); i++) {
lists.add(elections.get(i).getId() + " " + elections.get(i).getElection_name() + " " + elections.get(i).getActive());
System.out.println(elections.get(i).getId() + " " + elections.get(i).getElection_name() + " " + elections.get(i).getActive());
            }
```

```java
        }
        return lists;
    }
    @Autowired
    CandidatesRepository candidatesRepository;
    //增加候选人信息
    @PostMapping(value = "/candidates", params = {"eName","candidate"})
    public String addcandidates(String eName,String candidate) {
        Candidates candidates = new Candidates(eName,candidate);
        candidatesRepository.save(candidates);
        System.out.println("新增候选人的信息: " + candidates.toString() + ".");
        return "新增候选人的信息: " + candidates.toString() + ".";
    }
    @GetMapping(value = "/candidates")
    public List<String> queryAllCandidates() {
        List<String> lists = new LinkedList<String>();
        lists.add("所有候选人的信息: ");
        System.out.println("所有候选人的信息: ");
        List<Candidates> candidateLists = candidatesRepository.findAll();
        if (!candidateLists.isEmpty()) {
            for (int i = 0; i < candidateLists.size(); i++) {
                lists.add(candidateLists.get(i).toString());
                System.out.println(candidateLists.get(i).toString());
            }
        }
        return lists;
    }
    @Autowired
    private VotersRepository votersRepository;
    //增加选民信息
    @PostMapping(value = "/voters",params = {"vname","pubkey","signature","electionName"})
    public String addvoters ( String vname, String pubkey, String signature, String electionName) {
        Voters voters = new Voters(vname,pubkey,signature,electionName);
        votersRepository.save(voters);
        System.out.println("新增选民的信息: " + voters.toString() + ".");
        return "新增选民的信息: " + voters.toString() + ".";
    }
    @GetMapping(value = "/voters")
    public List<String> queryAllVoters() {
        List<String> lists = new LinkedList<String>();
        List<Voters> votersLists = votersRepository.findAll();
        lists.add("所有选民的信息: ");
        System.out.println("所有选民的信息: ");
        if (!votersLists.isEmpty()) {
            for (int i = 0; i < votersLists.size(); i++) {
                lists.add(votersLists.get(i).toString());
                System.out.println(votersLists.get(i).toString());
            }
```

```
        }
        return lists;
    }
}
```

10.3.10 创建类 VoteController

在包 com.bookcode.controller 中创建类 VoteController,代码如例 10-32 所示。

【例 10-32】 类 VoteController 的代码示例。

```java
package com.bookcode.Controller;
import com.bookcode.dao.*;
import com.bookcode.entity.*;
import org.springframework.beans.factory.annotation.Autowired;
import org.springframework.stereotype.Controller;
import org.springframework.web.bind.annotation.GetMapping;
import org.springframework.web.bind.annotation.PostMapping;
import org.springframework.web.bind.annotation.ResponseBody;
import java.util.LinkedList;
import java.util.List;
@Controller
public class VoteController {
    @Autowired
    private VotesRepository votesRepository;
    @Autowired
    private ElectionsRepository electionsRepository;
    @Autowired
    CandidatesRepository candidatesRepository ;
    @Autowired
    private VotersRepository votersRepository;
    @Autowired
    private ElectionBlockRepository electionBlockRepository;
    //登记选票
@PostMapping(value = "/votes",params = {"vname","pubkey","signature","electionName",
"candidate"})
public String addvotes(String vname, String pubkey, String signature, String electionName,
String candidate) {
        if( checkVoteInfo(vname,pubkey,signature,electionName,candidate)){
            System.out.println("此次投票无效,请重新投票.");
            return "/add_vote";
        }
        if (! checkBlockChainInfo(vname,pubkey,signature,electionName,candidate)){
            System.out.println("重复投票无效,请重新投票.");
            return "/add_vote";
        }
        Votes votes = new Votes(vname,pubkey,signature,electionName,candidate);
        System.out.println("您的投票信息: " + votes.toString() + ".");
```

```java
        votesRepository.save(votes);
        List<ElectionBlock> electionBlockList = electionBlockRepository.findAll();
        int blockno = 0;
        String strpreviousBlockHash = "-1";
        if(!electionBlockList.isEmpty()) {
            blockno = electionBlockList.get(electionBlockList.size()-1).getNo() + 1;
            strpreviousBlockHash = electionBlockList.get(electionBlockList.size()-1).getThisBlockHash();
        }
        long currentTime = System.currentTimeMillis();
        String strcurrentTime = String.valueOf(currentTime);
        String voteContent = vname + " " + pubkey + " " + electionName + " " + candidate + " " + strcurrentTime;
        String strthisBlockSimpleHash = String.valueOf(voteContent.hashCode());
        ElectionBlock electionBlock = new ElectionBlock(blockno, strcurrentTime, voteContent, strthisBlockSimpleHash, strpreviousBlockHash);
        System.out.println("您的投票区块链信息:" + electionBlock.toString() + ".");
        electionBlockRepository.save(electionBlock);
        return "/index";
    }
    //判断是否已经投票,若没有投则为 true
    private boolean checkBlockChainInfo(String vname, String pubkey, String signature, String electionName, String candidate) {
        int count = 0;
        List<ElectionBlock> electionBlockList = electionBlockRepository.findAll();
        if(electionBlockList.isEmpty()) {
            return true;
        }
        else{
            for(int j = 0;j<electionBlockList.size();j++){
                String voteContentChild = vname + " " + pubkey + " " + electionName + " " + candidate;
                if (0 == electionBlockList.get(j).getContent().indexOf(voteContentChild)){
                    count += 1;
                    break;
                }
            }
            if (count > 0)
                return false;
            else
                return true;
        }
    }
    //false 为无效
    private boolean checkVoteInfo(String vname, String pubkey, String signature, String electionName, String candidate) {
        int count = 0;
        //判断选举是否有效
        List<Elections> electionsList = electionsRepository.findAll();
        if(electionsList.isEmpty()) {
```

```java
            return false;
        }else{
            for(int j = 0;j < electionsList.size();j++){
                if(electionName.equals(electionsList.get(j).getElection_name())){
                    count += 1;
                    break;
                }
            }
            if (0 == count) return false;
        }
        //判断候选人是否有效
        List < Candidates > candidatesList = candidatesRepository.findAll();
        if(candidatesList.isEmpty()) {
            return false;
        }else{
            for(int i = 0;i < candidatesList.size();i++){
                if( electionName.equals(candidatesList.get(i).getElectionName()) &&
                        candidate.equals(candidatesList.get(i).getCandidateName())
                ){
                    count += 1;
                    break;
                }
            }
            if (1 == count) return false;
        }
        //判断选民是否有效
        List < Voters > votersList = votersRepository.findAll();
        if(votersList.isEmpty()) {
            return false;
        }else{
            for(int k = 0;k < votersList.size();k++){
                if( electionName.equals(votersList.get(k).getElectionName()) &&
                        vname.equals(votersList.get(k).getVname()) &&
                        pubkey.equals(votersList.get(k).getPublic_key()) &&
                        signature.equals(votersList.get(k).getPrivate_key())
                ){
                    count += 1;
                    break;
                }
            }
            if (2 == count) return false;
        }
        if(3 == count )
            return true;
        else
            return false;
    }
    @GetMapping(value = "/votes")
    @ResponseBody
    public List < String > queryAllVotes() {
```

```java
        List<String> lists = new LinkedList<String>();
        lists.add("所有投票的信息:");
        System.out.println("所有投票的信息:");
        List<Votes> votesLists = votesRepository.findAll();
        if (!votesLists.isEmpty()) {
            for (int i = 0; i < votesLists.size(); i++) {
                lists.add(votesLists.get(i).toString());
                System.out.println(votesLists.get(i).toString());
            }
        }
        return lists;
    }
    @GetMapping(value = "/blockchains")
    @ResponseBody
    public List<String> queryAllBlockchains() {
        List<String> lists = new LinkedList<String>();
        lists.add("所有区块链的信息:");
        System.out.println("所有区块链的信息:");
        List<ElectionBlock> electionBlockList = electionBlockRepository.findAll();
        if (!electionBlockList.isEmpty()) {
            for (int i = 0; i < electionBlockList.size(); i++) {
                lists.add(electionBlockList.get(i).toString());
                System.out.println(electionBlockList.get(i).toString());
            }
        }
        return lists;
    }
}
```

10.3.11 创建文件 index.html

在目录 src/main/resources/templates 下，创建文件 index.html，代码如例 10-33 所示。

【例 10-33】 文件 index.html 的代码示例。

```html
<!DOCTYPE html>
<html lang="en" xmlns:th="http://www.w3.org/1999/html">
<head>
    <meta charset="UTF-8">
    <title>简易投票系统首页</title>
</head>
<body>
<a th:href="@{/addelection}">增加选举项目</a><br>
<a th:href="@{/addvoters}">增加选民信息</a><br>
<a th:href="@{/addcandidates}">增加候选人信息</a><br>
<a th:href="@{/addvote}">登记投票信息</a><br>
<a th:href="@{/viewvotes}">查看投票结果</a><br>
</body>
</html>
```

10.3.12 创建文件 add_election.html

在目录 src/main/resources/templates 下,创建文件 add_election.html,代码如例 10-34 所示。

【例 10-34】 文件 add_election.html 的代码示例。

```html
<!DOCTYPE html>
<html xmlns:th = "http://www.w3.org/1999/xhtml">
<head>
    <title>发起选举页面</title>
    <meta http-equiv = "Content-Type" content = "text/html; charset = UTF-8">
</head>
<body>
<div>发起一项选举</div>
<form name = "addelection" th:action = "@{/elections}" method = "post">
    选举的名称：<input type = "text" name = "name" th:value = "${name}"><br/>
    <input type = "submit" value = "发起" />
</form>
</body>
</html>
```

10.3.13 创建文件 add_voters.html

在目录 src/main/resources/templates 下,创建文件 add_voters.html,代码如例 10-35 所示。

【例 10-35】 文件 add_voters.html 的代码示例。

```html
<!DOCTYPE html>
<html xmlns:th = "http://www.w3.org/1999/xhtml">
<head>
    <title>增加选民信息页面</title>
    <meta http-equiv = "Content-Type" content = "text/html; charset = UTF-8">
</head>
<body>
<h3>选民信息登记</h3>
<p>选民必须注册后才能在特定选举中投票.为了注册,选民的个人信息必须与要密钥一起提供.
</p>
<form name = "addvote" th:action = "@{/voters}" method = "post">
    Voter Name(选民姓名)：<input type = "text" name = "vname" th:value = "${vname}"><br/>
    Public Key(公钥)：<input type = "text" name = "pubkey" th:value = "${pubkey}"><br/>
    Private Key(私钥)：<input type = "text" name = "signature" th:value = "${signature}">
    <br/>
    Election(选举)：<input type = "text" name = "electionName" th:value = "${electionName}">
    <br/>
```

```
           <input type="submit" value="增加" />
    </form>
  </body>
</html>
```

10.3.14 创建文件 add_vote.html

在目录 src/main/resources/templates 下,创建文件 add_vote.html,代码如例 10-36 所示。

【例 10-36】 文件 add_vote.html 代码示例。

```
<!DOCTYPE html>
<html xmlns:th="http://www.w3.org/1999/xhtml">
<head>
    <title>投票页面</title>
    <meta http-equiv="Content-Type" content="text/html; charset=UTF-8">
</head>
<body>
<h3>选民投票信息登记</h3>
<p>选民必须注册后才能在特定选举中投票.为了注册,选民的个人信息必须与要密钥一起提供.
</p>
<form name="addvote" th:action="@{/votes}" method="post">
    Voter Name(选民姓名): <input type="text" name="vname" th:value="${vname}"><br/>
    Public Key(公钥): <input type="text" name="pubkey" th:value="${pubkey}"><br/>
    Private Key(私钥): <input type="text" name="signature" th:value="${signature}">
<br/>
    Election(选举): <input type="text" name="electionName" th:value="${electionName}">
<br/>
    Candidate(候选人): <input type="text" name="candidate" th:value="${candidate}">
<br/>
           <input type="submit" value="投票" />
    </form>
  </body>
</html>
```

10.3.15 创建文件 add_candidates.html

在目录 src/main/resources/templates 下,创建文件 add_candidates.html,代码如例 10-37 所示。

【例 10-37】 文件 add_candidates.html 的代码示例。

```
<!DOCTYPE html>
<html xmlns:th="http://www.w3.org/1999/xhtml">
<head>
```

```html
        <title>增加候选人信息页面</title>
        <meta http-equiv="Content-Type" content="text/html; charset=UTF-8">
</head>
<body>
<h3>候选人信息登记</h3>
<form name="addcandidates" th:action="@{/candidates}" method="post">
        Election(选举):<input type="text" name="eName" th:value=" ${eName}"><br/>
        Candidate(候选人):<input type="text" name="candidate" th:value=" ${candidate}">
<br/>
        <input type="submit" value="增加" />
</form>
</body>
</html>
```

10.3.16 创建文件 view_votes.html

在目录 src/main/resources/templates 下,创建文件 view_votes.html,代码如例 10-38 所示。

【例 10-38】 文件 view_votes.html 的代码示例。

```html
<!DOCTYPE html>
<html xmlns:th="http://www.w3.org/1999/xhtml">
<head>
        <title>查看投票结果页面</title>
        <meta http-equiv="Content-Type" content="text/html; charset=UTF-8">
</head>
<body>
<div>投票结果</div>
<form name="blockchains" th:action="@{/blockchains}" method="get">
        <input type="submit" value="查看区块链(投票)详细信息" />
</form>
</body>
</html>
```

10.3.17 修改配置文件 application.properties

修改配置文件 application.properties,修改后的代码如例 10-23 所示。

10.3.18 运行程序

运行程序,在浏览器中输入 localhost:8080,结果如图 10-14 所示。单击图 10-14 中"增加选举项目"链接,结果如图 10-15 所示。在图 10-15 中输入选举名称(be1)后单击"发起"按钮,返回增加的选举项目信息,如图 10-16 所示。单击图 10-14 中"增加选民信息"链接,结果如图 10-17 所示。在图 10-17 中输入选民信息后单击"增加"按钮,返回增加的选民信息,如图 10-18 所示。单击图 10-14 中"增加候选人信息"链接,结果如图 10-19 所示。在图 10-19 中输入候选人信息后单击"增加"按钮,返回增加的候选人信息,如图 10-20 所示。单击图 10-14 中"登记投票信息"链接,结果如图 10-21 所示。在图 10-21 中输入投票信息后单击"投票"按钮,控制台中输出增加的投票信息,如图 10-22 所示。单击图 10-14 中"查看投票结果"链接,结果如图 10-23 所示。单击图 10-23 中"查看区块链(投票)详细信息"按钮,结果如图 10-24 所示。

图 10-14　在浏览器中输入 localhost:8080 的结果

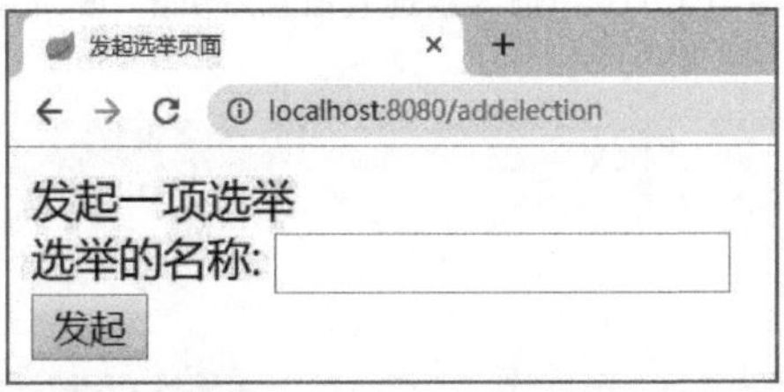

图 10-15　单击图 10-14 中"增加选举项目"链接的结果

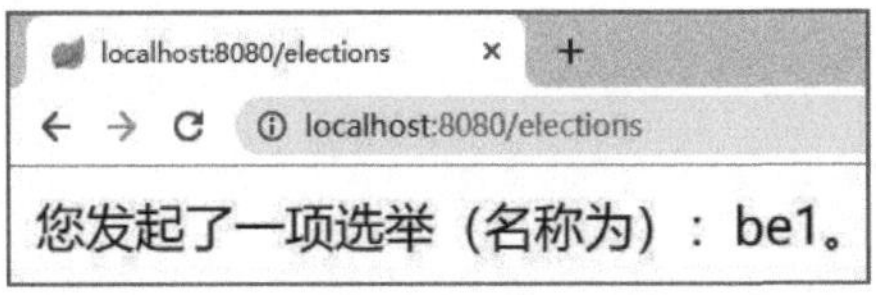

图 10-16　在图 10-15 中输入选举名称后单击"发起"按钮的结果

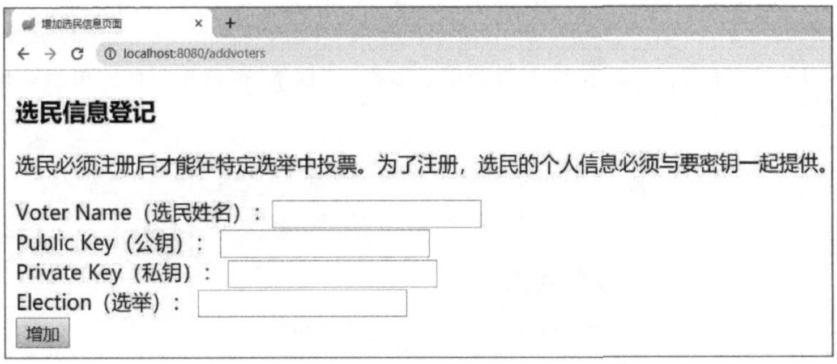

图 10-17　单击图 10-14 中"增加选民信息"链接的结果

新增选民的信息：Voters(id=55, vname=zsf, electionName=zsf, public_key=fsz, private_key=be1)。

图 10-18　在图 10-17 中输入选民信息后单击"增加"按钮的结果

候选人信息登记

Election（选举）：
Candidate（候选人）：
[增加]

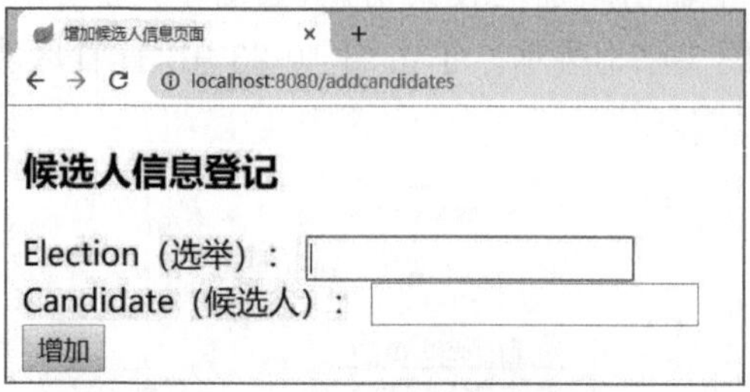

图 10-19　单击图 10-14 中"增加候选人信息"链接的结果

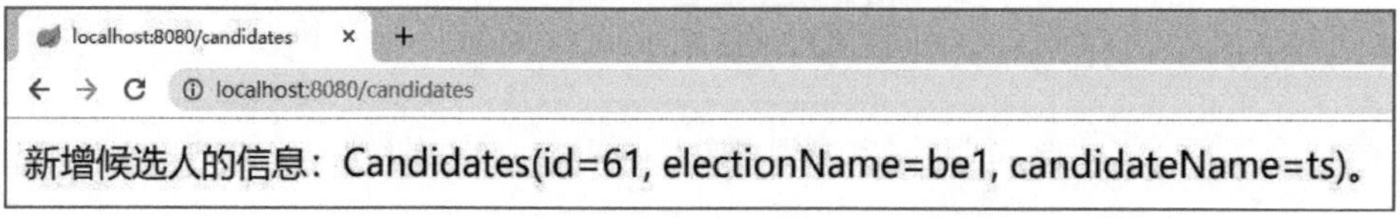

新增候选人的信息：Candidates(id=61, electionName=be1, candidateName=ts)。

图 10-20　在图 10-19 中输入候选人信息后单击"增加"按钮的结果

选民投票信息登记

选民必须注册后才能在特定选举中投票。为了注册，选民的个人信息必须与要密钥一起提供。

Voter Name（选民姓名）：
Public Key（公钥）：
Private Key（私钥）：
Election（选举）：
Candidate（候选人）：
[投票]

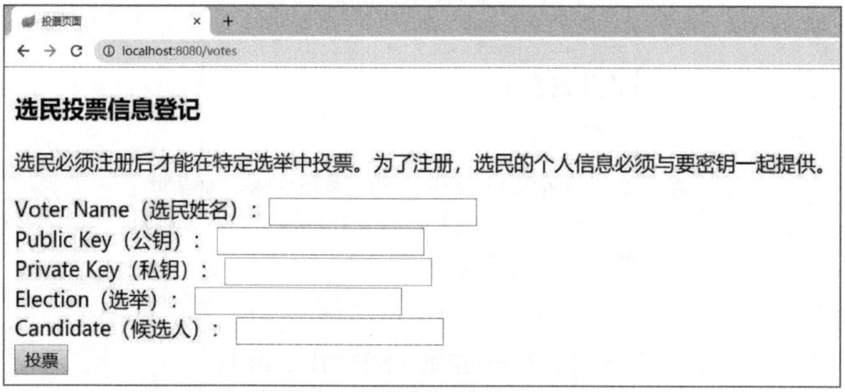

图 10-21　单击图 10-14 中"登记投票信息"链接的结果

您的投票信息：Votes(id=null, vname=sdf, pubkey=sdf, prikey=sdf, election_name=et2, candicate=c1)。
您的投票区块链信息：ElectionBlock(id=null, no=11, epochTime=1561591021636, content=sdf sdf et2 c1 1561591021636,

图 10-22　在图 10-21 中输入投票信息后单击"投票"按钮控制台的输出结果

投票结果
[查看区块链（投票）详细信息]

图 10-23　单击图 10-14 中"查看投票结果"链接的结果

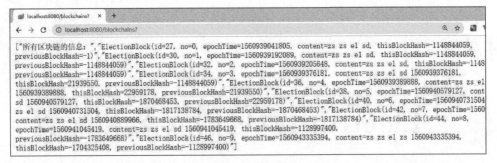

图 10-24　单击图 10-23 中"查看区块链（投票）详细信息"按钮后的结果

习题 10

实验题

请独立完成书上的案例。

附录A

Electrum钱包的安装和配置

本附录介绍 PC 版和网页版 Electrum 钱包的安装和配置。

A.1　PC 版 Electrum 钱包的安装和配置

先从官方网站(https://www.electrum.org/#home)上下载可执行文件,然后双击可执行文件,显示安装路径,如图 A-1 所示。单击 Install 按钮进行安装,安装过程中的界面如图 A-2 所示。

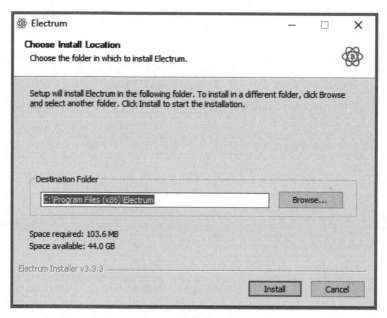

图 A-1　选择安装路径

附录A　Electrum钱包的安装和配置

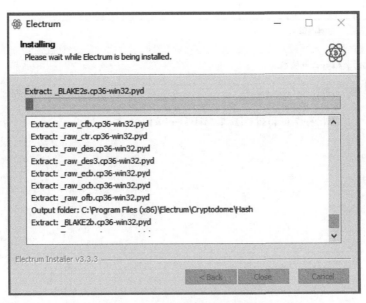

图 A-2　安装过程中的界面

安装完成后，单击 Close 按钮，计算机桌面上出现 Electrum 小图标。双击小图标后，可以按照"安装向导"逐步配置钱包信息。第一次配置图标时，直接选择所有默认项，并单击"下一步"按钮即可，如图 A-3～图 A-10 所示。注意，在第六步时要记下密语种子，第七步要输入该密语种子，第八步输入自己的密码，在以后登录时要用到该密码。

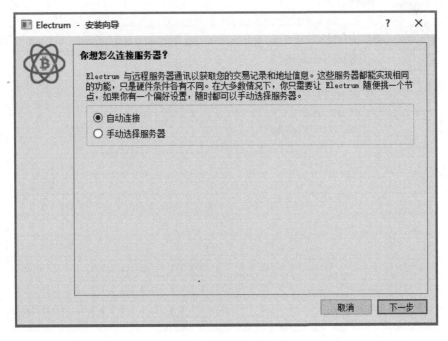

图 A-3　配置钱包第一步

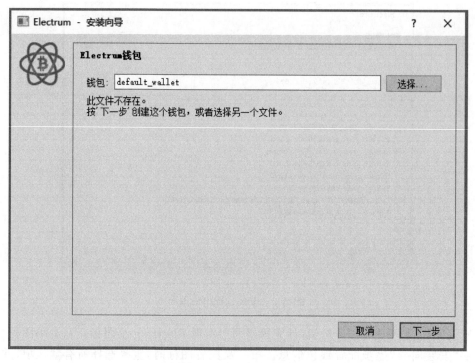

图 A-4　配置钱包第二步

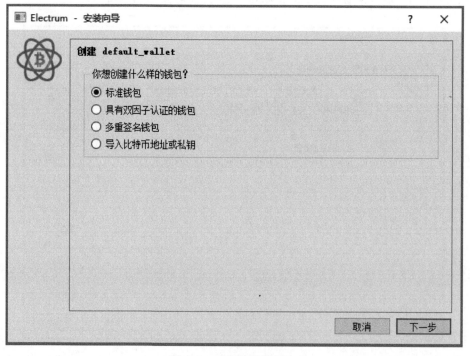

图 A-5　配置钱包第三步

附录A　Electrum钱包的安装和配置

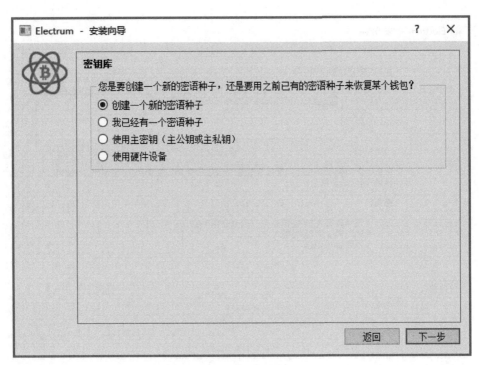

图 A-6　配置钱包第四步

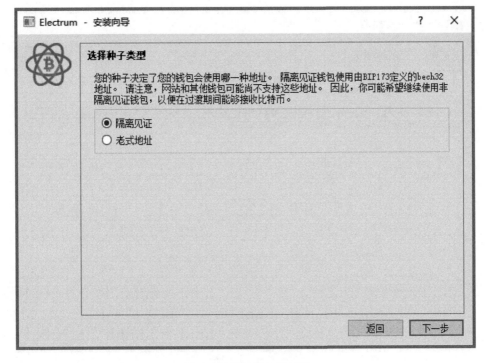

图 A-7　配置钱包第五步

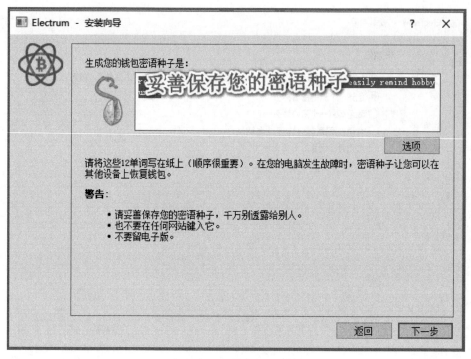

图 A-8 配置钱包第六步

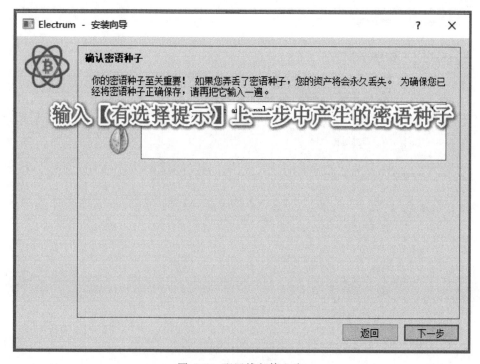

图 A-9 配置钱包第七步

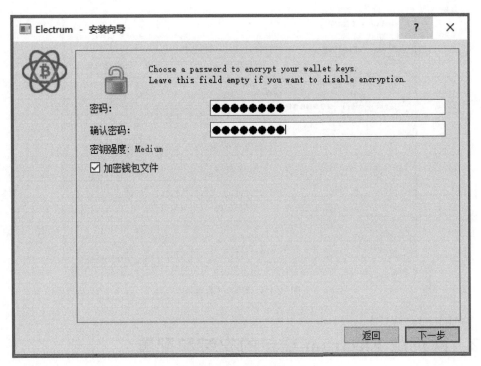

图 A-10　配置钱包第八步

配置完成钱包信息后，自动弹出对话框，以供用户选择是否自动检查钱包更新，如图 A-11 所示，单击 Yes 按钮即可。

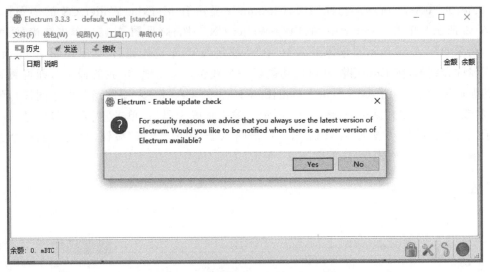

图 A-11　选择是否自动检查钱包更新

下一次登录已有钱包时，仍然是双击计算机桌面上的 Electrum 小图标，出现如图 A-12 所示的"安装向导"，输入在第八步中设置的密码即可登录钱包。

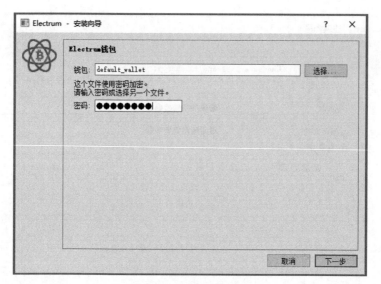

图 A-12　登录已有钱包

A.2　网页版 Electrum 钱包的安装和配置

　　单击 MyEtherWallet 官方网站(https://www.myetherwallet.com/)首页中的链接 MyEtherWallet CX,跳转到谷歌浏览器(Chrome)的添加扩展程序页面,如图 A-13 所示。单击 Add to Chrome 按钮后,将扩展程序 MyEtherWallet CX 成功安装到谷歌浏览器 Chrome,如图 A-14 所示。与此同时,Chrome 中显示有 MyEtherWallet CX 的小图标,如图 A-15 所示。单击 Chrome 中 MyEtherWallet CX 小图标(见图 A-15),单击链接"添加钱包",结果如图 A-16 所示。选择图 A-16 中"生成新钱包"单选按钮后,结果如图 A-17 所示。输入昵称和密码,按 Enter 键,即可成功创建一个钱包。单击链接"我的钱包",即可看到钱包的昵称和地址,如图 A-18 所示。单击图 A-18 中的 view 链接(眼睛形状的小图标),弹出"查看钱包信息: a"对话框,如图 A-19 所示。输入正确密码后,单击"查看钱包信息"按钮,显示钱包的相关信息,如图 A-20 所示。单击图 A-20 中的"下载"按钮,自动生成一个关于钱包信息的 JSON 格式文件,如图 A-21 所示。

图 A-13　单击 MyEtherWallet 官方网站首页中链接 MyEtherWallet CX 的结果

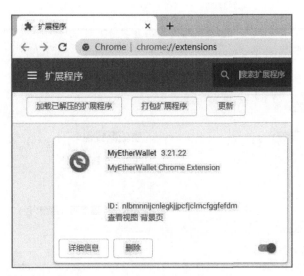

图 A-14　成功将扩展程序 MyEtherWallet CX 安装到谷歌浏览器 Chrome 的结果

图 A-15　成功将扩展程序 MyEtherWallet CX 安装到谷歌浏览器 Chrome 后浏览器中的小图标

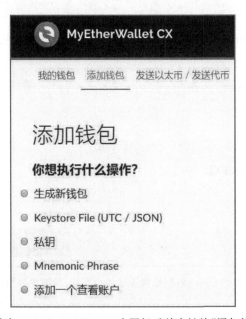

图 A-16　单击 MyEtherWallet CX 小图标后单击链接"添加钱包"的结果

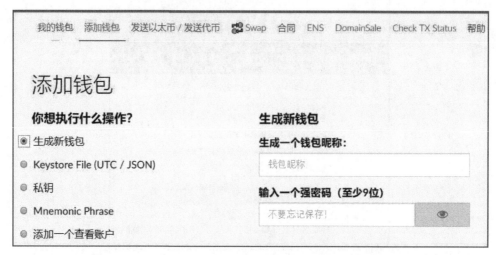

图 A-17　选择图 A-16 中"生成新钱包"单选按钮后的结果

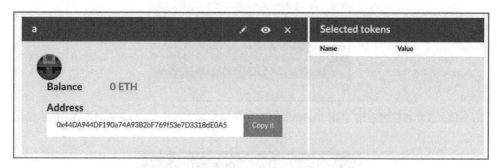

图 A-18　成功创建钱包后单击链接"我的钱包",即可看到钱包的昵称和地址

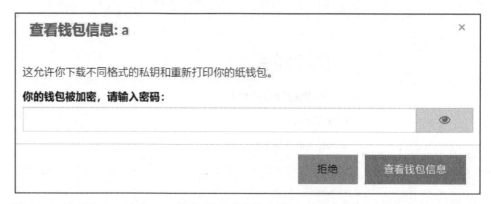

图 A-19　单击图 A-18 中的 view 链接(眼睛形状的小图标)后弹出的"查看钱包信息：a"对话框

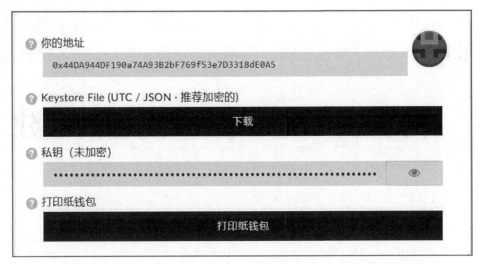

图 A-20　输入正确密码后单击"查看钱包信息"按钮,显示钱包的相关信息

图 A-21　单击图 A-20 中的"下载"按钮后自动生成一个关于钱包信息的 JSON 格式文件的结果

附录B 网页版Coin.Space钱包的创建

本附录主要介绍网页版Coin.Space钱包的创建和设置。

登录官方网站(https://coin.space),单击"新建电子钱包"按钮,如图B-1所示。在随后的界面中单击"生成安全种子"按钮,如图B-2所示。

图B-1 单击"新建电子钱包"按钮

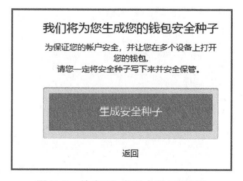

图B-2 单击"生成安全种子"按钮

单击Copy to Clipboard链接,复制种子,复制成功后Copy to Clipboard链接会变成Copied;妥善保存好安全种子之后,选择"我已经将安全种子写下来或以其他方式安全地妥善存储"和I Agree To The Terms & Conditions复选框,单击"设置您的登录口令"按钮,如图B-3所示。设置登录口令,如图B-4所示。

下次登录官方网站后,输入登录口令即可,如图B-5所示。

图 B-3 保存"生成安全种子"

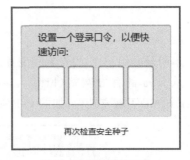

图 B-4 设置登录口令

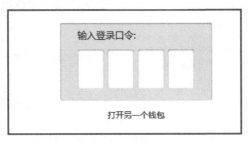

图 B-5 已有钱包输入登录口令

附录C JDK的安装和配置

本附录介绍 JDK 的安装和配置。

从 Java 的官方网站（http://www.oracle.com/technetwork/java/javase/downloads/index.html）下载安装包，如图 C-1 所示。

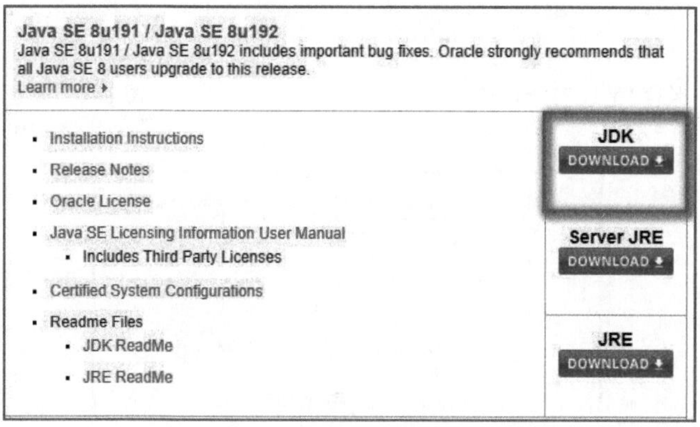

图 C-1 从官方网站下载 JDK 安装包

安装完成后，设置系统变量 JAVA_HOME，如图 C-2 所示。配置好 JAVA_HOME 之后，将 %JAVA_HOME%\bin 加入系统的环境变量 path 中，如图 C-3 所示。

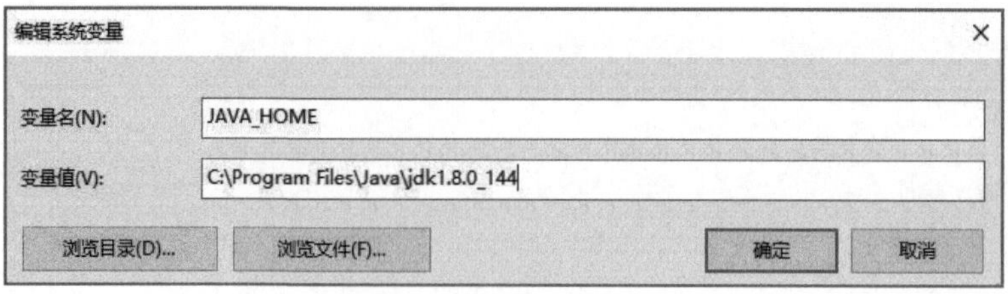

图 C-2 设置系统变量 JAVA_HOME

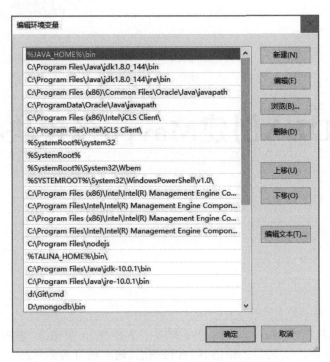

图 C-3　设置环境变量 path 中 JDK 的路径信息

附录D IDEA创建Maven多模块项目

创建新项目时,选择 Maven,如图 D-1 所示。单击 Next 按钮后,输入项目的 GroupId、ArtifactId 信息,如图 D-2 所示。单击 Next 按钮后,输入项目名称和路径,也保留默认的项目名称和路径,如图 D-3 所示。删除项目的 src 文件夹,如图 D-4 所示。通过在项目 pom.xml 文件中增加如例 D-1 所示的代码,修改项目的打包方式,如图 D-5 所示。

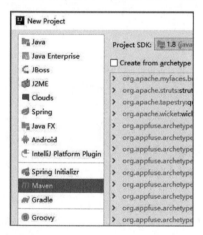

图 D-1　创建新项目时选择 Maven

图 D-2　输入项目的 GroupId、ArtifactId

图 D-3　设置新项目的名称和路径

附录D　IDEA创建Maven多模块项目

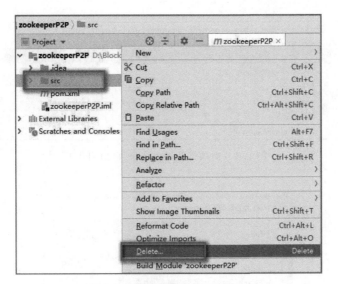

图 D-4　删除项目的 src 文件夹

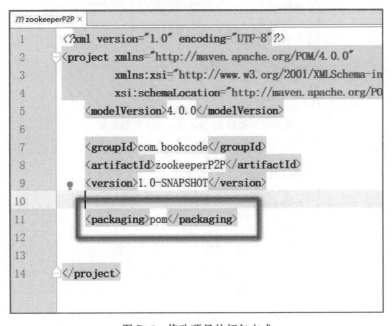

图 D-5　修改项目的打包方式

【例 D-1】修改项目打包方式的代码示例。

```
<packaging>pom</packaging>
```

右击项目名称后选择 New 菜单的 Module 子菜单，向项目增加新模块（Module），如图 D-6 所示。选择 Spring Initializr 类型模块，如图 D-7 所示。输入模块的 Group、Artifact 等信息，如图 D-8 所示。单击 Next 按钮后，选择新模块的依赖，如图 D-9 所示。

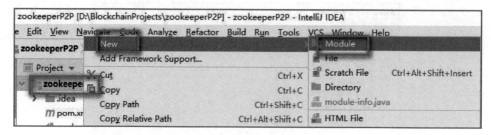

图 D-6　向项目新增一个模块（Module）

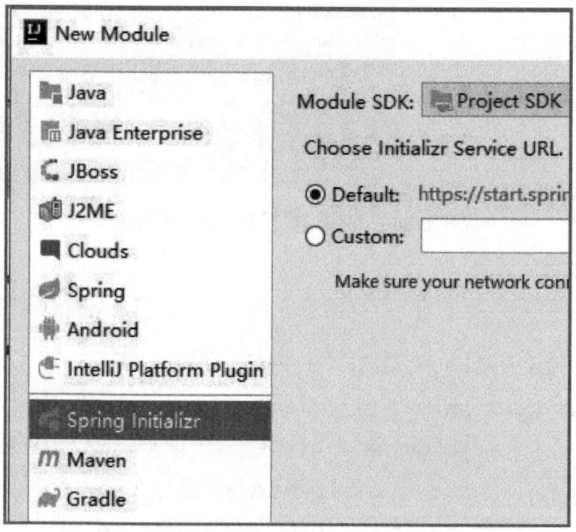

图 D-7　选择 Spring Initializr 类型模块

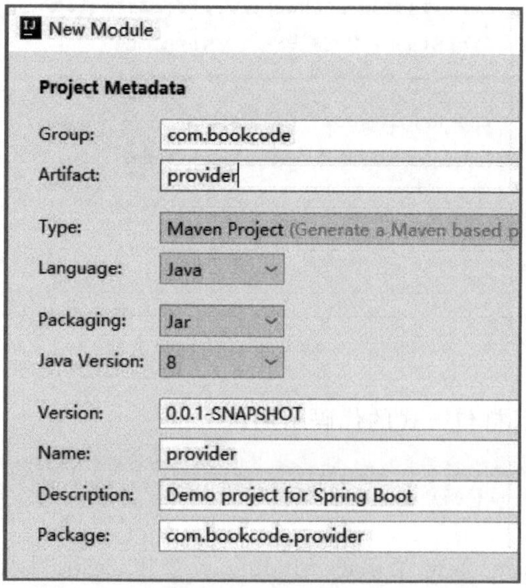

图 D-8　输入模块 Group、Artifact 等信息

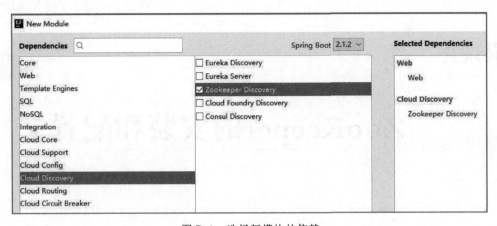

图 D-9　选择新模块的依赖

附录 E

ZooKeeper的安装和配置

从 Apache Zookeeper 的官方网站（https://archive.apache.org/dist/zookeeper/）或国内镜像网址（如 http://mirrors.hust.edu.cn/apache/zookeeper/）下载压缩包，如图 E-1 所示。

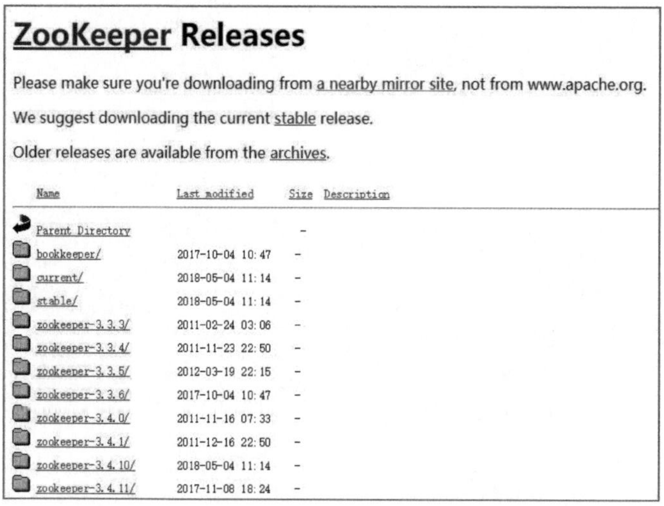

图 E-1　从官方网站下载 ZooKeeper 压缩包

解压后，在目录 zookeeper 下，建立 data、log 两个文件夹，如图 E-2 所示。在目录 zookeeper 的子目录 conf 下创建文件 zoo.cfg，如图 E-3 所示。文件 zoo.cfg 的内容和文件 zoo_sample.cfg 的内容相同。在此基础上向文件 zoo.cfg 中增加 data 和 log 的文件夹信息，增加文件夹信息的代码如示例 E-1 所示。

【例 E-1】　向文件 zoo.cfg 增加 data 和 log 文件夹信息的代码示例。

```
dataDir = D:\zookeeper\data
dataLogDir = D:\zookeeper\log
```

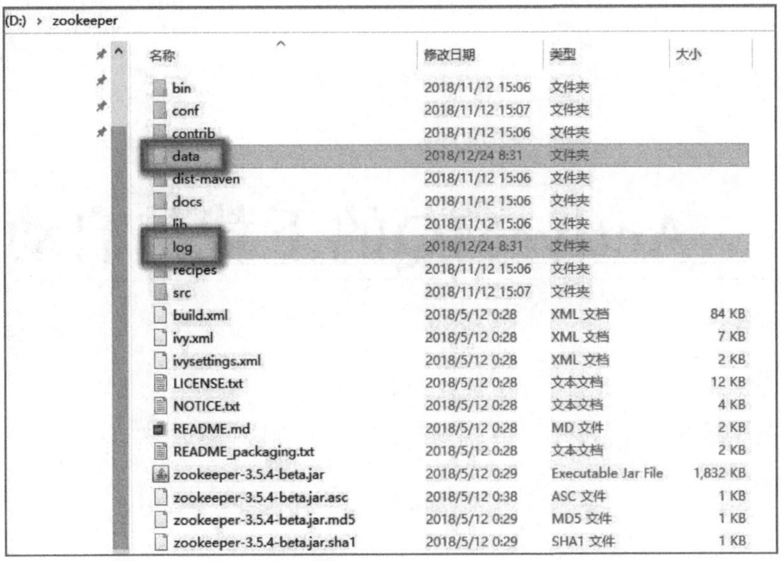

图 E-2　解压 Apache Zookeeper 压缩包后在目录 zookeeper 下新建 data 和 log 文件夹

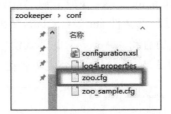

图 E-3　在目录 zookeeper 的子目录 conf 文件夹下新增 zoo.cfg

将 Apache Zookeeper 的路径信息加入系统的环境变量 path 中，如图 E-4 所示。

图 E-4　设置系统变量 path 中 Apache Zookeeper 的路径信息

附录F ActiveMQ的下载与启动

从官方网站(http://activemq.apache.org/)下载 ActiveMQ 最新的二进制版本,解压缩后如图 F-1 所示。双击 bin 目录下文件 activemq.bat,结果如图 F-2 所示。图 F-2 中矩形框内 started 表示 ActiveMQ 成功启动。ActiveMQ 默认使用的 TCP 连接端口是 61616,通过如例 F-1 所示的命令,可以测试 ActiveMQ 是否成功启动,结果如图 F-3 所示。

图 F-1　ActiveMQ 解压缩后的文件夹

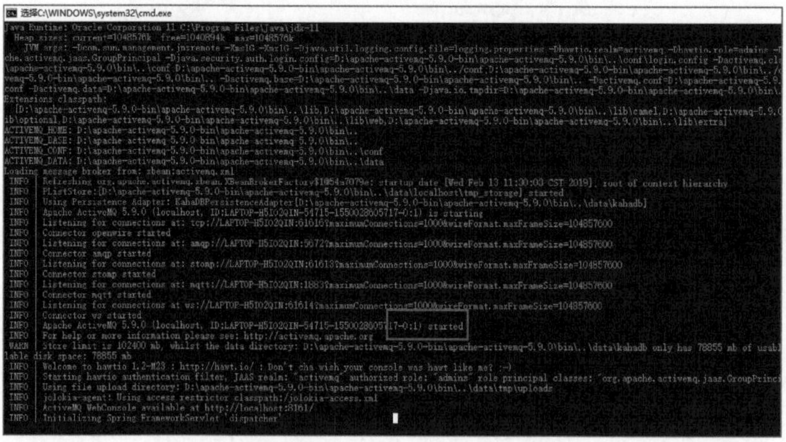

图 F-2　ActiveMQ 启动效果

图 F-3 对 ActiveMQ 启动与否的测试

【例 F-1】 测试 ActiveMQ 是否成功启动的命令示例。

```
netstat -an|find "61616"
```

附录G

RabbitMQ的安装与配置

RabbitMQ 是用 Erlang 语言编写的,安装 RabbitMQ 的前提是安装 Erlang。从官方网站上下载 Erlang 安装文件后,双击可执行文件,单击 Next 按钮,选择一个用于安装 Erlang 的目录,然后依次单击 Next 按钮、Finish 按钮就可以完成 Erlang 的安装。安装完成后,设置系统变量 ERLANG_HOME,如图 G-1 所示。将%ERLANG_HOME%\bin 加入环境变量 path 中。

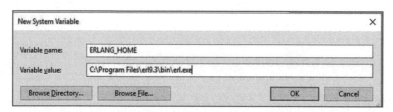

图 G-1 设置系统变量 ERLANG_HOME

从官方网站下载 RabbitMQ 最新版本 rabbitmq-server-3.7.7.exe,然后进行安装。

用如例 G-1 所示的命令安装 rabbitmq_management 插件,安装这款插件后可以以可视化的方式来操作 RabbitMQ。执行完命令后,若看到如图 G-2 所示的结果就表示插件安装成功了。

【例 G-1】 安装 rabbitmq_management 插件的命令示例。

```
rabbitmq-plugins enable rabbitmq_management
```

可以用如例 G-2 所示的命令启动 RabbitMQ 服务器。

【例 G-2】 启动 RabbitMQ 服务器的命令示例。

```
rabbitmq-server
```

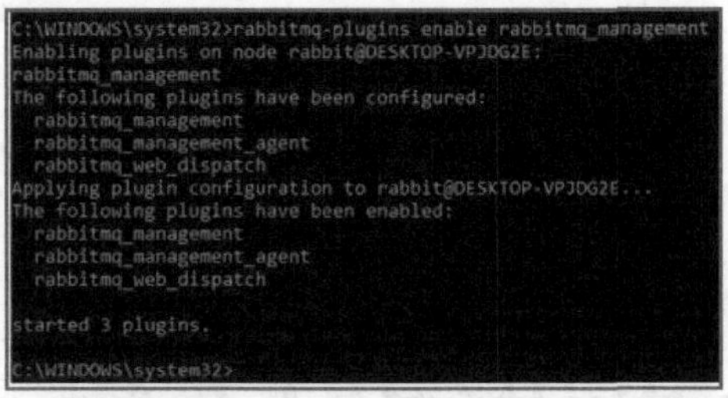

图 G-2 安装 rabbitmq_management 插件成功的效果

在浏览器中输入 http://localhost:15672 后可以看到如图 G-3 所示的登录界面，用户名和命名均为 guest。RabbitMQ 安装好后是作为 Windows 服务在后台运行的。

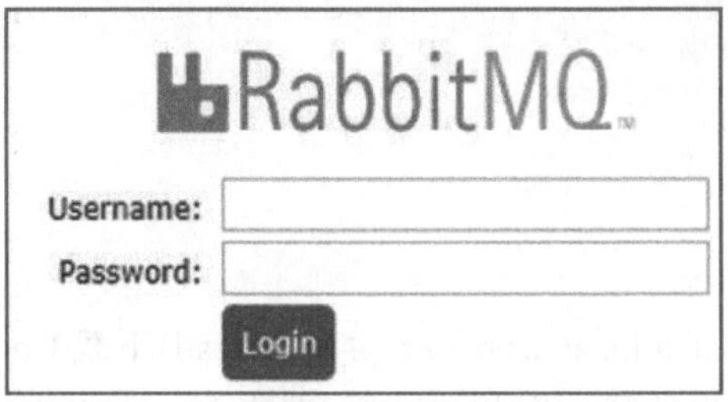

图 G-3 登录界面

附录H CouchDB的安装与配置

从官方网站(http://couchdb.apache.org/#download)下载 CouchDB 安装包,如图 H-1 所示。选择安装向导中每一步的默认安装选项后,单击 Next 按钮,如图 H-2 所示。安装成功后,在 Windows 命令行程序中执行完如例 H-1 所示的命令后,再在浏览器中输入 http://127.0.0.1:5984/_utils/,结果如图 H-3 所示。

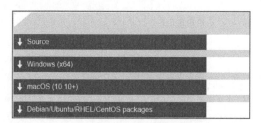

图 H-1　CouchDB 下载页面

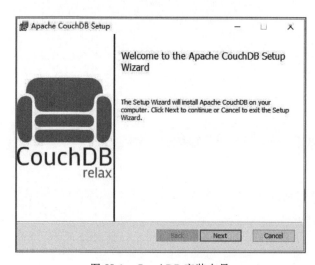

图 H-2　CouchDB 安装向导

图 H-3 安装成功后在浏览器中输入 http://127.0.0.1:5984/_utils/ 的结果

【例 H-1】 启动 CouchDB 数据库的命令示例。

```
couchdb
```

参 考 文 献

[1] 牛冬.区块链底层设计Java实战[M].北京：电子工业出版社,2019.
[2] 武源文,柏罡,温江凌.区块链：交易系统开发指南[M].北京：电子工业出版社,2019.
[3] 朱志文.Node.js区块链开发[M].北京：机械工业出版社,2017.
[4] 张增骏,董宁,朱轩彤,等.深度探索区块链：Hyperledger技术与应用[M].北京：机械工业出版社,2018.
[5] 曹源,张翀,丁兆云,等.DAG区块链技术：原理与实践[M].北京：机械工业出版社,2018.
[6] 朱建明,高胜,段美姣,等.区块链技术与应用[M].北京：机械工业出版社,2018.
[7] 陈东敏.世界因区块链而不同[M].北京：北京航空航天大学出版社,2017.
[8] 安德烈亚斯·安东诺普洛斯.区块链：通往资产数字化之路[M].林华,等译.北京：中信出版社,2018.
[9] 郑天民.微服务设计原理与架构[M].北京：人民邮电出版社,2018.
[10] 闫洪磊.Activiti实战[M].北京：机械工业出版社,2015.
[11] 杨熳.基于区块链技术的会计模式浅探[J].新会计,2017(9)：57-58.
[12] 司淑娴.大数据时代对会计行业的重塑——基于区块链视角的分析[J].财会研究,2017(9)：24-28.
[13] DAI J,VASARHELYI M A. Toward Blockchain-Based Accounting and Assurance[J]. Journal of information systems,2017,31(3)：5-21.
[14] 钟玮,贾英姿.区块链技术在会计中的应用展望[J].会计之友,2016(17)：122-125.
[15] 黄步添,刘琦,何钦铭,等.基于语义嵌入模型与交易信息的智能合约自动分类系统[J].自动化学报,2017,43(09)：1532-1543.
[16] ANTONOPOULOS A M. Mastering Bitcoin：Programming the open blockchain[M]. Sebastopol：O'Reilly Media,Inc.,2017.
[17] MASCIANDARO D. Central Bank Digital Cash and Cryptocurrencies：Insights from a New Baumol-Friedman Demand for Money[J]. Australian Economic Review,2018,51(4)：540-550.

图书资源支持

感谢您一直以来对清华版图书的支持和爱护。为了配合本书的使用,本书提供配套的资源,有需求的读者请扫描下方的"书圈"微信公众号二维码,在图书专区下载,也可以拨打电话或发送电子邮件咨询。

如果您在使用本书的过程中遇到了什么问题,或者有相关图书出版计划,也请您发邮件告诉我们,以便我们更好地为您服务。

我们的联系方式:

地　　址:北京市海淀区双清路学研大厦 A 座 701

邮　　编:100084

电　　话:010-83470236　010-83470237

资源下载:http://www.tup.com.cn

客服邮箱:2301891038@qq.com

QQ:2301891038(请写明您的单位和姓名)

资源下载、样书申请

书 圈

扫一扫,获取最新目录

课 程 直 播

用微信扫一扫右边的二维码,即可关注清华大学出版社公众号"书圈"。